I0113948

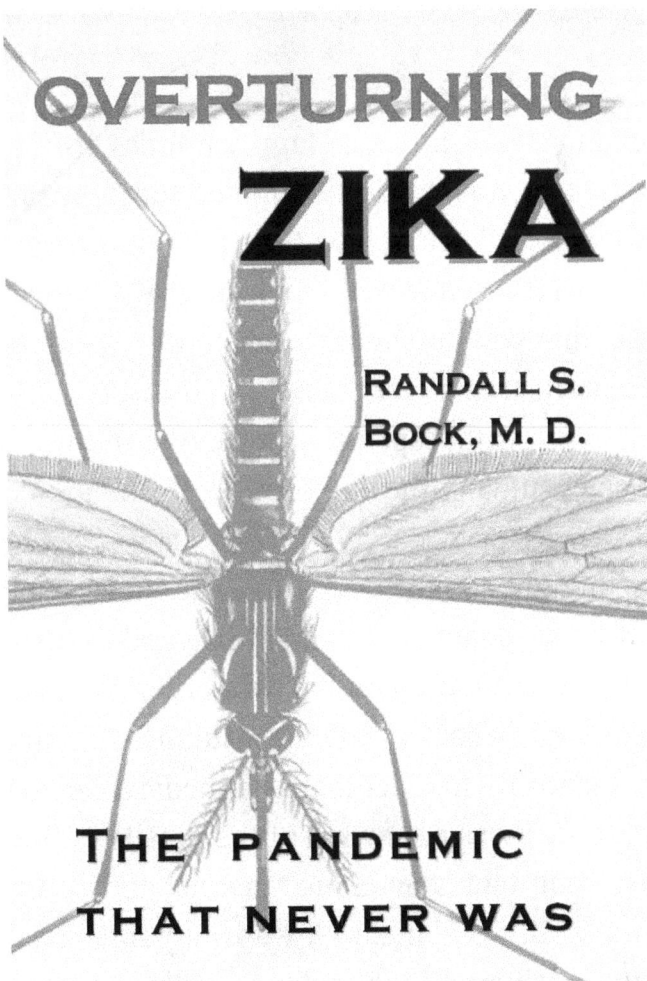

OVERTURNING
ZIKA

RANDALL S.
BOCK, M. D.

THE PANDEMIC
THAT NEVER WAS

Foreword

<u>The ZIKA VIRUS Pandemic Mystery</u>: "In 2016 we were planning our next Christmas family holiday for 2017 and South America was the destination of choice. Brazil was the country we all decided was the no.1 destination to visit, but there was one problem - the ZIKA-virus pandemic that was "ravaging" that country held unknown threats that most members of the family were uncomfortable with, especially those of childbearing age. Chile was eventually the compromise destination chosen.

"I never thought about ZIKA after that until by some stroke of luck, in late 2021 when I listened to Randall Bock's live presentation on the mystery of the pandemic that never was. What I heard was astounding -- and I was more disturbed by the fact that as a medical doctor and a Chief of ICU, I was not aware of the details of this last and most important "pandemic" that happened before COVID-19. How this material was not published before, and why it's taken so long for it to get the air-time it deserves -- could be the biggest errors committed before the COVID-19 pandemic.

"This book will alter how you interpret the public health information you receive through mainstream media. It is a piece of medical history that should be recommended reading for all major public health decision-makers; the journalism, medical, and economic fraternities -- essentially everyone who is impacted by, and cares how narrative-driven decision-making (based on weakly constructed facts not openly challenged) can derail public health and destroy a country's socioeconomic structure, far beyond that particular incident. It is also a lesson to the world: if unheeded, these smaller incidents may otherwise eventually lead to catastrophic interventions, worldwide and poorly-implemented!"

Dr. Nathi Mdladla --Intensive Care Cardiac anesthetist, Johannesburg, South Africa [1]

"Overturning Zika is a timely and relevant exposé of the 2015 Zika-microcephaly pandemic, which fueled fear and panic the world over, before disappearing into apparent oblivion mere months later. Equal parts mystery and scientific fact, **Overturning Zika** is a fascinating, engaging and often frightening account of the power-plays behind modern medicine, and the agendas pursued, seemingly in the interests of public health.

"This book is a must-read. You'll learn more about medicine, scientific process and politics than you ever thought possible, and all in one go. I can highly recommend it, not only to those in the medical and scientific fields, but also to those who simply want their eyes opened to the realities of a world in which "all is not what it seems"."

Heidi Short -- Statistician, Pandata.org[2]

"With **Overturning Zika** Randall Bock puts relevant historical perspective around the COVID-19 hysteria. Far too many well-intentioned people—on both sides of any debate around COVID-19—have fallen into the same trap: *"This is __new__, and all discussions about it are therefore about THESE data, THIS virus, and THIS scenario."*

"Although the Zika panic was confined largely to Brazil, many, if not most of the signs were there for what careful observers noticed relatively early on about the COVID-19 panic.
- Identification of a novel virus without legitimate scientific proof of its novelty? *"Check"*
- Sound-bite promotion of a narrative without legitimate and independent verification? *"Check"*

- Use of the public square as a means of promotion of a specific hypothesis? *"Check"*
- The use of the panic to enhance the careers of specific scientists? *"Check"*

"While one could **assume** that similar techniques were used before, Bock's analysis provides pointed, specific, obvious parallels. He proves that these techniques were used, almost word-for-word! For anyone who might be inclined to say, *"Never again!"*, while thinking the COVID-19 narrative was materially different, Bock's analysis offers a hearty, *"Are you new?"* **Overturning Zika** provides a sobering review of techniques and outcomes that the COVID-19 hysteria simply recycled. Only this time, it was worldwide!"

Wilton D. Alston Senior Safety Engineer, TÜV Rheinland North America

"Courageous and powerfully executed", this is a vast underestimation of the achievement of Randy Bock M.D. in **Overturning Zika**; a thorough and systematic exploration of the (somewhat shaky) evidence behind the momentary Zika pandemic. Books of this genre are unbelievably important at this time in our history and will form the basis of texts of importance in holding the public health establishment to account in the future.

"The manipulative, secondary misuse of the Zika phenomenon: to effect political ends -- easily meets the test of what is considered dishonorable. Randy lays out his evidence and chronology in a format that enables a clear understanding of the facts and acts as the foundation to explain and understand later institutional overreaction to Covid-19. With truth, we can make a change -- for that reason, this is an essential read."

Cherry Hughes, MSc; Advanced Clinical Practice

"If not for the mosquitos, one might, at times, think this eye-opening read was referring to Covid-19 instead of Zika. The parallels in the public health response between Covid-19 and Zika are remarkable -- from inconsistencies in 'case' -definitions; to diagnoses based on subjective data; to censorship of alternative thinking.

"And did you think the Zika "pandemic" was over and no longer relevant? Our government continues to fund multiple investigational vaccines for Zika, some already in Phase 2. Like Covid-19, Zika is considered a medical countermeasure (MCM)-related issue:

> "FDA stands ready to use our authorities to the fullest extent to help facilitate the development and availability of products for Zika virus. Under the FDA's Emergency Use Authorization (EUA) mechanism...." [3]

Here we go again?"

Carol Taccetta, MD,[4] FCAP, 25-year drug-development physician

"Given all that the "*experts*" got wrong with the COVID-19 pandemic, it's important to ask, "*what else have they gotten wrong?*" With **Overturning Zika**, physician Randy Bock lights a fuse under a powder keg of questions that threaten to blow up the entire official narrative of the Zika pandemic, from the unknown number of Zika cases to their unsubstantiated role in microcephaly, while exposing the politicized and sloppy science that provoked unnecessary fear and suffering in millions of expectant mothers in tropical regions worldwide."

Steve Templeton, Ph.D.: Immunologist; Author of *FEAR OF A MICROBIAL PLANET* [5]

Title pages

Overturning Zika

The Pandemic That Never Was

by Randall S. Bock, M. D.

Drivestraight Publishing

Cover design: David Amyx, Showrav Dev

ISBN: 978-1-957975-01-6 FIRST EDITION Paperback

Bock, Randall: **Overturning Zika** *"The Pandemic That Never Was"*; Publisher: Drivestraight Inc.

randybock.com

Acknowledgments:

This book is the result of a years-long process. As such there are many people to whom I owe my gratitude. First and foremost are my late and beloved parents, Samuel and Dorothy Bock, without whose love and support nothing would have been possible, let alone this book. They sacrificed material comfort so their sons could have the best educations. As a five-year-old, I came up with a silly song "*If You Think You Can Do It, You Can*", but Mom didn't think it was silly, and always brought me around to remember it. My father had my back when a speech I'd written (as a teenager) proved quite controversial. Their trust became the strongest grounding for a life of ideas.

I also have to thank the book's earliest readers and supporters: Seth Corey, MD; Errol Lincoln Uys; Bharani Padmanabhan, MD, PhD; Terry Russell, PhD; Young Jae Koh; Robin Kaczmarczyk; Neto Lócio; Roy Wallen; Susan Patton, Michael Hurd, PhD, Brent Kievit-Kylar, Steven Coogan; Chris and Domini Gordon– and many others from Pandata.org. My efforts would have waned without their encouragement and diminished without their clarity.

Most of all, I owe limitless gratitude to my wife, Lynne LeBlanc, who has stuck with me through thick and thin –

and has made every aspect of my life fuller, better, and more rewarding. She has been a patient listener, thoughtful reader and piquant editor. I appreciate her love, generosity, and the styling of each of our days, making certain the otherwise mundane sparkles.

The idea for **Overturning Zika** long preceded the Covid-19 pandemic's recurring lockdowns and diminishments of social spheres; nonetheless, one personal silver-lining of this era was the focus and concentration those lent towards this book's production. "*How the time flew*" when I was semi-isolated, camping out (hours on end) ensconced within our upstairs' deck: amidst birds' songs and occasional gentle winds (weather permitting); unforeseen benefits of a very-much unwanted retrenchment's homeward, inward turn.

Why read "Overturning Zika"?

Overturning Zika is an exposé of the Zika-microcephaly phenomenon. Zika rose abruptly from conjecture to panic. After enduring briefly as a worldwide pandemic, Zika mysteriously disappeared -- with only silence from health officials and experts.

Billions whipsawed by the recent Covid-19 threat have the inadvertent background necessary to delve into Zika's own wild story. **Overturning Zika** represents a *"Get Out of Jail Free"* card for 100-million pregnant women in the tropics, many still self-sequestering in a misapplied fear of any single given mosquito bite.

Overturning Zika intertwines personal stories of intrepid, mission-driven physicians and researchers, with clarifying thumbnails of background science. The Zika story has the excitement of the 1990s movie *Twister:* Brazil's captivating rogue physicians on a quest to find new mysterious illnesses -- but with a Tower of Babel result: an insufficiently-proven pandemic's misuse and misdirection for personal and political ends. Brazil's adventurers will ultimately have harmed more than helped: creating a 'tornado' whose damage through misinformation will exceed that from a mosquito's infection.

It was a dramatic whirlwind of events in Brazil 2015 that led to redefinition of previously harmless Zika – and then (separately) to a claim of more microcephaly cases, elsewhere. It is shocking, in retrospect, to find literally zero Zika-tests had been performed, and that microcephaly was neither firmly defined nor previously counted in Brazil. Both claims were suspect, and the connection between the two even more so.

Overturning Zika is timely and ultimately inevitable. It is on the "right side of history" – although 'history' may not yet realize it. As the years go by with the continued absence of Zika microcephaly, this exposé will stand as a beacon of forthright amendment and remedy both to Zika -- and to the mechanism of medical science that allowed it to pass unchecked into public consciousness, bringing hysteria, fear and despair. The hundreds of thousands of children not among us, their conceptions or deliveries annulled due to unfounded Zika mandates can't ask for this book, but future parents will and should.

About the author —
Randall S. Bock, M. D. is a Yale and University of Rochester educated primary care physician. Even without direct experience with Zika or the tropics, it seemed to Dr. Bock rather an anomalous situation, medically: that dengue's biologic twin (Zika) caused a problem (microcephaly) never previously diagnosed from

dengue's millions of cases, endemically in Brazil. His pursuit led to a Zika-review published in 2020 not followed up by the medical, scientific, or journalistic community – in part (as expressed in personal conversation with the editors of JAMA) so as not to cast doubt on the public health establishment. Further efforts have produced this book and a commentary piece in the American Journal of Medicine, July 2022. [6]

Table of Contents

Contents

Foreword   2

Title pages   8

Acknowledgments: 10

Why read "Overturning Zika"?   12

   About the author —   13

Table of Contents   15

PREFACE and dedication   17

Synopsis   24

   An even shorter synopsis: 32

Introduction: like a comet, Zika burned brightly, then disappeared   33

Zika, A Shock to The System   39

Zika, dengue: *"separated at birth"*?   43

"Medicine is a social science, ...   53

"Haste makes waste" vs. "Too little too late"?   63

The Eureka Moment, Microcephaly in Recife   69

The Family Business   73

Bridging the New Illness to a New Problem   82

Intermission   87

Checks and Balances?   93

The Shifting Sands of Shifting Standards   108

The Microcephaly Bubble   117

Mosquitoes Know No Borders... The Net Result?   125

   Between states within Brazil: 135

Between Brazil and other tropical countries within the region: 137

Viral Penetration, Rubella versus Zika   141

A closer look at Brazil's Zika-microcephaly epicenter, Recife Brazil   149

Hypothesis Testing: When Results Fail Beliefs   159

Never Let a Crisis Go to Waste   172

Your Bodies, Our Choice; Zika's Key Unlocking Abortion   172

Cheerleading Team-Zika   179

Theories of Zika's Disappearance Abound   187

The PRESENCE of an ABSENCE   187

The Current State of Zika Research   190

OBJECTION: 192

"Your Honor, the question assumes facts not in evidence."   192

"Yeah! That's the Ticket!"   197

The Experts Respond   205

The Next Zika pandemic? Coming Soon!?   206

Summary of the academic Zika-expert's Q&A   212

A Shorter Expert Exchange   215

The Broken Promise of the Zika Vaccine; Women's Lives in Limbo   217

What Is the Opposite of "*Warp Speed*"?   217

The Project Sputters   219

Infect People to Save Them?   224

Coding Coda   231

The Verdict   233

Overturning Zika's Entirely Different "*Mission*":   244

INDEX   254

NOTES, CITATIONS, REFERENCES   260

PREFACE and dedication

This book is dedicated to young mothers (and fathers) all around the world, the providers of new life, the direct passage to love and respect for their children, our next and newest generation. They have lived the last six years under fear that a single mosquito bite could irrevocably harm their pregnancies. Zika-microcephaly has essentially disappeared throughout the world, but with none of the fanfare of its arrival. May this book provide young, expecting couples relief and clarity; less fear, more knowledge.

The last previous "great pandemic" was 2015-2016's Zika-microcephaly, whose underlying premise was that:

> *A virus (Zika) heretofore completely harmless caused a terrifying congenital abnormality which none of its near-identical viral "cousins" (like dengue) ever had.*

Almost as mysteriously as it arrived, it disappeared, from the news, from the clinics, from consciousness. But there has been no Zika -update or -downgrade within medicine, science, public health, Wikipedia or the news, since -- despite the fact that ...

> *"It is the nature of any inquiry that more information is gained with the passage of time.*

Today's information is more complete than yesterday's. It affords better insight, deeper perspective and sounder judgment."[7] Holman Jenkins, Jr.

Overturning Zika intends not just to remember Zika, but to reclassify it – and return it to its prior role of medical near-anonymity, albeit with the eternal footnote of having brought panic and distress to the world's populace, and thought-conformity (with so very little in the way of critical review) to its journalistic, medical, and scientific classes.

Before delving into the story, please consider this personal disclaimer: as a physician, my medical expertise is in primary care; not tropical disease or neuropediatrics. I have never seen or treated Zika illness nor congenital microcephaly. I am neither medical researcher nor Brazil-authority; however, much as the Wizard of Oz needed no witchcraft (but only careful observation) to realize and interpret Lion's bravery; Scarecrow's smarts; Tin Woodman's heart – I have evaluated Zika-microcephaly's events and evidence; studies and results by the standards of the scientific method– and made logical determinations. As the days, months, and years continue (fortunately) to pass without any replication of the originally claimed Zika-microcephaly connection – its underlying theory seems ever more doubtful...

Is the Zika microcephaly theory valid if its results were never repeated anywhere else or at any other time after the original in 2015 NE Brazil?

With each and every year following the original claims of a Zika microcephaly pandemic continuing not to demonstrate any increase in microcephaly, worldwide, the theory seems ever more improbable. So this magic eight ball says "VERY DOUBTFUL", all signs point to "no".

Yet, there have been no retractions from the scientists, institutions, universities, health-ministries, and public health organizations (like the CDC and WHO) central to raising alarms in 2016 -- and as a result, no relief for the convinced-as-vulnerable one billion women of

reproductive age within the tropics who continue to worry
that one errant mosquito-bite will irreversibly damage
their cherished hope within. Imagine having that
additional pressure during the most precious phase of a
couple's life, bringing a new loved one into existence.

There is no shame in having been swept away at the time,
concerned over any potential inadequacy in response to
the announced Zika-pandemic emergency; however, one
is reminded of the Churchillian[8] concept that a healthy
skepticism coincides with greater experience and
familiarity of history.

People ask me "Why are *you* writing a book on Zika?" I
was interested (but puzzled) when I first encountered the
topic in the news early 2016, believing that within this
strange new disease connection there was something
medically and logically amiss. Any conclusive answers,
though, would have to wait until the illness ran its course
one way or another. Three years later, prodded by a
spontaneous "Facebook Memory" reposting of my query,
I became curious once again.

Randy Bock
January 30, 2016 · 🔔

Oddly, #WSJ page-one #ZIKA story's twins: only one has microcephaly.

Disproving congenital-exposure theory?

There is a lot about which to be curious. Listeners are stunned that throughout 2015 there were no clinical Zika-tests available anywhere in Brazil, that the entire theory evolved without a single confirmed Zika-case, in real-time. Furthermore, there was no Brazilian microcephaly registry available for determining an increase, no firm diagnostic standard; no verification in retrospect; and no increase the following year.

Late 2019, I completed an 800-word review, thinking it would be a slam dunk to have it published, as it potentially overturned the Zika-microcephaly paradigm. Nonetheless, *"Lessons from the Burst Zika Bubble"*[9] was rejected by every major medical and science journal and news outlet. American Greatness took a chance on it in March 2020, but by then Covid had arrived.

Certain medical journal editors and science writers felt it was "no longer the right time" to bring up Zika, fearing that my report might diminish trust in Covid's public health initiatives. This ignores that the audience (the "public" in "public health") comprises thinking adults who understand the importance of directives during an emergency but who also value the respect given through reconsideration. Trust in our institutions is enhanced knowing they are resilient and confident enough not to be threatened by fallibility.

A "wrong" was committed on tens of millions of Brazilians, young women starting families – particularly those poor less able to avoid *Aedes aegypti* mosquitoes. Moreover the poor and less educated remain less likely to receive information on the apparent disappearance of Zika and/or Zika-microcephaly in the interim. Physicians and researchers originating this series of conjectures didn't knowingly take advantage of the population; however, they have not reasonably circled back to undo the alarm. Instead, the theory is continually propped up,

with claims of variant strains, hidden immunity, dengue cross-reactivity.

Overturning Zika grew to cover a problem larger than Zika: the often-haphazard decision process surrounding what is claimed as "science" -- and then fixed in the imagination (later to become operative in the political sphere) as "knowledge". Public health decisions have huge ramifications. If they are to work best for the people, there must be a process of quality assurance, self-observation and reevaluation. Sudden disruptions in the balance between risk-avoidance and the exercise of freedom opens a pathway to grave secondary problems: excess opioid deaths after Covid; diminished births after Zika. Similarly, resources devoted without firm foundation (e.g. to Zika) invariably result in shortfalls elsewhere.

Synopsis

a shortcut to events in the book (or skip to chapter 1):

- Zika (discovered in Uganda in 1947) until 2015 had never been claimed as responsible for any human illness. Zika had never prior appeared in the Americas. Moreover, there were no routine clinical Zika-testing capabilities anywhere on earth at the time of Brazil's announcement.

- In the mid-2000s there was a bit of a "buzz" about Zika as certain dengue cases in the Pacific were relabeled (at a far remove) by the CDC.

- Zika was "discovered" to be in Brazil April 2015, by semi-independent physicians (Drs. Carlos Brito and Kleber Luz) and researchers (Drs. Silvia Sardi and Gubio Soares Campos, "**S&SC**"). They determined and attributed Zika to a set of milder dengue-patients and others with achiness and rash. There were no laboratory-test- or clinical confirmations on any specific patients, in real time.

- **S&SC**'s research laboratory's Zika PCR-test "primers" on hand, had been left by a visiting Senegalese researcher the year prior never to be confirmed by Brazil's "FDA" or other researchers for efficacy or accuracy.

- Zika and dengue are physically and genomically nearly identical, thus cross-reactive in the lab.

Notably, dengue is endemic to the areas studied and has never caused congenital microcephaly.

- There was pushback from institutional researchers' pointing out flaws in the Zika claims, but those were quashed through **S&SC'**s leaking to the popular press, creating a story that took on its own life, independent of review.

- Medical fame (/notoriety) attaches to those who are able to find NEW DISEASES. Careers were made, including **S&SC'**s.

- Completely separately, microcephaly was declared to be "epidemic" by a small group of neuropediatricians in Recife, WITHOUT referencing any prior-year microcephaly-registry data, without engaging institutional research study or support.

- 2015-Brazil's erstwhile absence of a national microcephaly registry precluded any comparison-data review; moreover even if there had been one, it would have been relatively meaningless given the complete absence of consistent diagnostic standards for microcephaly. Commonly acceptable international standards had not yet been adopted. Shortly after the declared epidemic, standards were made much more stringent.

- Recife's Dr. Carlos Brito was brought in by the Ministry of Health to evaluate the microcephaly situation and quickly associated the locally perceived increase to Zika (which if this were business could be seen as a "conflict of interest").

- Dr. Brito asked microcephaly-birth mothers (and apparently not any control group) retroactively about rashes, aches they MIGHT have had months prior, during first trimester pregnancy, determining those as Zika-cases, without laboratory confirmation.

- No Zika lab tests were available in Brazil clinically in 2015

- The Zika-microcephaly connection also was leaked to the press at a time that there were skeptical institutional reviews recorded and in process. The news took on a life of its own and very shortly became a worldwide scare, in part because of the coincidence of the 2016 Olympics' being in Rio de Janeiro.

- Microcephaly is a rare syndrome not attributable to one specific cause except for the very severe genetic version which is not applicable in the Zika-microcephaly saga.

- Confirmed cases of microcephaly (months later) ultimately comprised fewer than 5% of the original panic-era claims implying physicians' overdiagnosis from a combination of panic, overcaution, and Brazil's incorrect and inconsistent microcephaly standards at the time. Brazil applied

stricter (and stricter still) microcephaly criteria twice DURING the presumed microcephaly epidemic, both times reducing the microcephaly overdiagnosis (false-positive) numbers.

- The following year when microcephaly-standards had been firmed up, and Zika-diagnoses were possible via laboratory, there was no increase in microcephaly seen.

- Microcephaly (as claimed) concentrated in and coincided with the location and timing of the news-generated panic (in Recife and Brazil's Northeast) rather than with the vector-mosquito's own range. This differs from (Zika-twin) dengue's illness, which occurs when and where its shared vector (Aedes aegypti mosquito) does.

- Even in Recife itself, microcephaly shows certain neighborhoods at orders of magnitude more than others, i.e. wealthy neighborhoods didn't exhibit microcephaly even though there was no reason to have been overly cautious regarding mosquitoes anywhere in the tropical city in the year prior. Arguably wealthy people have better mosquito nets - but they also have less exposure to CMV (cytomegalovirus) insecticides, alcoholism, drug use and other factors bringing on prematurity.

- Northeast Brazil is poor, ethnically shorter in stature, with far more prematurity and other ancillary maternal issues that track with microcephaly. The "Indios" of Recife are physically smaller than the ethnic Europeans from whom the microcephaly standards were developed from more European southern Brazil (Rio, São Paulo). So too are their babies.

- a confluence of interests helped shape and push the Zika-microcephaly narrative, shunning the necessary scientific re-examination the case warranted.

- The Zika-microcephaly connection's flame was fanned by social and political concerns aligned with its continuation; for instance as a means to reverse (Catholic) Brazil's abortion prohibitions.

- The embattled erstwhile impeached president Dilma Rousseff found an ally in the illness as a nationalistic rallying call, drawing troops up.

- Calls for complete stoppage of pregnancies (from certain health ministers and advocates) "Women needed to avoid pregnancy somehow. Because clearly nothing else was going to save their babies." (Zika: The Emerging Epidemic - Donald G. McNeil Jr.)

- Brazil's North is chronically (relatively) impoverished, darker-skinned, with lesser health infrastructure. Zika/microcephaly was seen by some as a tool by which to address "equity" issues. Money flowed in during the Zika emergency from

WHO, CDC and other countries' grants in research.

- There was also the tension between research institutions and official "science" versus the more "charismatic" (in both the religious and personal senses of the word) physicians who were pushing for in a sense the blessing of this rare problem - but one by which these other inequities and inequalities would be reversed.
- National funds award stipends to mothers with babies deemed microcephalic FROM erstwhile Zika.
- From the following year (2016) and onwards, with Zika-testing gradually coming on board, more refined and consistent microcephaly measurement and standards, and overall awareness - Zika-microcephaly DISAPPEARED as a phenomenon. It didn't recur in the presumed hotspot (Northeast Brazil), nor anywhere else on the globe.
- Retrospective data-modeling reconstructions cast doubt that there was any genuine increase in microcephaly in the first place. These studies seem not to have been widely broadcast to the public, nor adequately adopted within the scientific academy to force reconsiderations.
- Zika, when seen in other countries, has not brought microcephaly rate increases.

- Albert Einstein said, "*There is an age-old adage, 'If the facts don't fit the theory, change the theory.' But too often it's easier to keep the theory and change the facts.*"
- Zika-microcephaly's absence since the presumed pandemic hasn't brought scientists to question the validity of the underlying theory.
- Instead, reinforcing Albert Einstein's cynicism, numerous Band-Aids to the original theory are applied: for instance some theorize:
    - 2015-Brazil had a particularly dangerous "mutant strain" – or,
    - that 2015 alone brought immediate Zika herd-immunity to Brazil (never the case with its twin, dengue).
    - that the public health effort reversed Zika via awareness and avoidance.
- But none of these really hold water considering that every other tropical country avoided the Zika-microcephaly correlation, notwithstanding an absence of herd immunity to Zika in general or any "mutant strain" in particular. None had public health campaigns comparable to Brazil's.
- Research funding, once turned "*on*", is targeted to be kept "*on*" by the researchers themselves. No one admits errors. Obviously, there will be no retractions, no reformulations, despite the massive

ABSENCE of further supporting data in the years hence on epidemiologic, population basis.

- Zika may not be the biggest news anymore, but warnings about outdoor exposure are still given to prospective mothers in the United States and elsewhere around the world. It should be removed as a concept.

- The WHO, the CDC, health experts and epidemiologists are understandably effective during emergencies, but less so in self-correcting. Mandates can be enforced, but genuine trust requires work.

- There have been literally thousands of articles written on Zika post-2015, and seemingly none question the underlying premise(s).

- "SCIENCE" can be defined both as a body of knowledge, AND the refined, reproducible process of collecting and confirming that knowledge. Thus "questioning science" IS "science". Science has no official "court" to determine rulings; rather a free and open discussion mostly through journal articles -- although opposing views on this topic are not at all visible despite lack of the theory's holding up in the meanwhile. Science ultimately won't do well or be trusted if it embodies aspects more of a priesthood than a debate.

An even shorter synopsis:

The scientific method (for testing theories) has not been followed and respected. Very little was done for step #2.

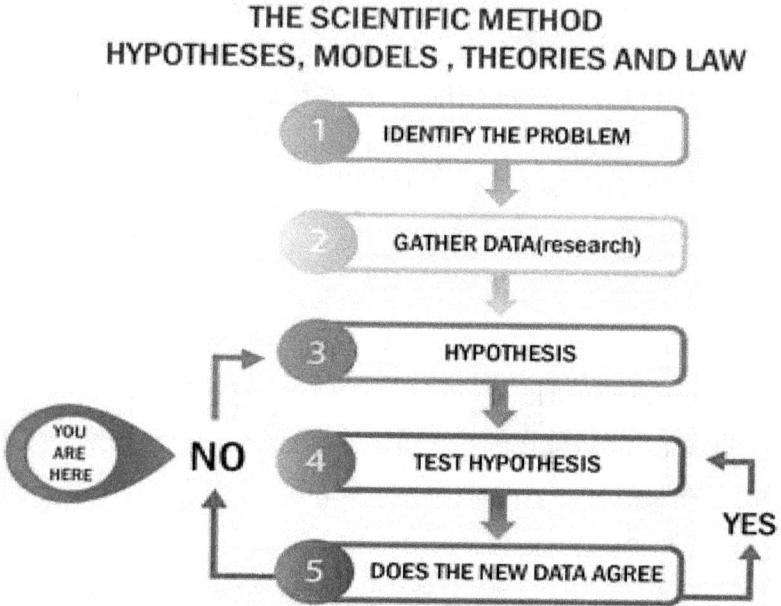

THE SCIENTIFIC METHOD
HYPOTHESES, MODELS , THEORIES AND LAW

1 IDENTIFY THE PROBLEM

2 GATHER DATA(research)

3 HYPOTHESIS

YOU ARE HERE   NO   4 TEST HYPOTHESIS   YES

5 DOES THE NEW DATA AGREE

Media leaks of conjectures led to panic and compromised the timing and ability to gather adequate data to enforce the Zika-microcephaly hypothesis. Experiments are difficult or impossible when dealing with severe human illnesses; nonetheless, when no recurrences or instances occurred at any other place or time, the hypothesis should have been rejected or reconfigured. *Overturning Zika* represents the explanation of why *"You Are Here"*, acknowledging that subsequent data have not agreed with the original premises and conjectures of Zika's causing microcephaly.

Introduction: like a comet, Zika burned brightly, then disappeared

The Zika virus burst from Brazil's tropics into the news, 2016, as a vast, new, and dangerous global health threat: announcing a pandemic of microcephaly births: neonates with small heads and diminished intellect. Previously, in the 60 years since its detection, Zika had been of no human consequence, known only by tropical disease specialists.

Instantly, upon the announcement of an unproven mosquito-Zika-microcephaly connection, Brazil was turned upside down: a panicked population stayed indoors; 220,000 Brazilian soldiers were diverted to public health measures including "door-to-door mosquito prevention"; travel advisories threatened derailment of the 2016 Rio Olympics; insect repellent disappeared from the shelves; and certain health officials suggested women avoid pregnancies indefinitely![10] [11]

A worldwide spread was anticipated to reach similarly tropical areas which in total host 40% of the world population, 3.3 billion people. With guesses of Zika's microcephaly rate conservatively at 2.3%,[12] Zika threatened more than a million microcephalic births, yearly -- with no vaccine, no treatment, and no end in sight![13] [14] [15] Within Brazil, government directives and general fear PREVENTED ~120,000 births, nearly the entirety of whom would have been normal children.[16] Billed (*by Dr. Anthony Fauci, February 2016*) as an "explosive pandemic",[17] Zika understandably dominated the news.

In the following years, there have been **no** Zika-microcephaly hotspots -- despite a well-funded global medical surveillance's finding Zika-seropositivity in every inhabited continent since. For instance, Zika was endemic in Rajasthan India 2018, but without any increase in microcephaly.[18] Much like a comet, Zika burned brightly, then disappeared from view.

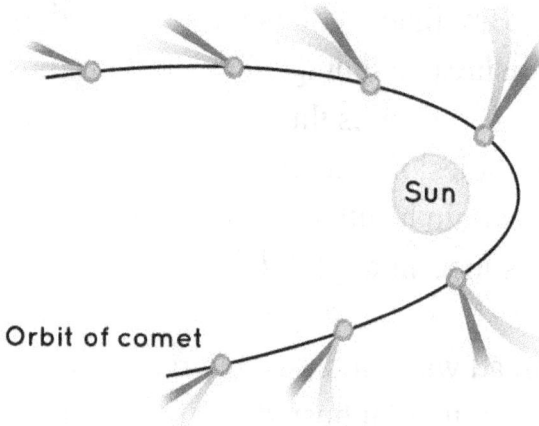

Nonetheless, nearly 2 full years into the continuing Zika-disappearance -- in December 2017, Dr. Fauci (with all of its import and implications for his NIAID) perhaps

inescapably continue to see Zika as more than perhaps it is or ever was.

The Journal of
Infectious Diseases

Pandemic Zika: A Formidable Challenge to Medicine and Public Health

David M Morens   Anthony S Fauci   December 2017

How have these words held up to the test of time?

> *"Finally, it is important that we not assume that pandemic Zika is a one-time crisis that can be met, controlled, and then forgotten or relegated to historical review. Over the past 4 decades, we have seen the dawn of a new infectious disease era.... All the tropical world and much of the temperate world is now at risk and is likely to remain at risk for the foreseeable future. How we deal with the Zika pandemic is likely to become a roadmap for future challenges."* [19]

Whatever is *the opposite of "pandemic"* is where Zika-microcephaly has actually remained in every subsequent year: absent from public discourse, public health news, and fortunately, the public itself.

What is the opposite of pandemic?

abnormal, exclusive, incomplete, infrequent, rare, uncommon, unusual local, discrete, separate, single, nonuniversal, independent

Nonetheless, Dr. Fauci, the CDC and WHO have not backtracked from Zika as a menace. Now, of course, "pandemic" interest has shifted to COVID-19, overshadowing "ZIKV-15". Is it time yet to disagree with Dr. Fauci and presume that "pandemic Zika (actually) *IS* a one-time crisis", ready to be "relegated to historical review"?

Even as *Overturning Zika* answers, "yes", still we should not forget Zika's vast personal, political, and business disruptions– what Dr. Fauci blithely recommended as the Zika pandemic's *"roadmap for future challenges."* Indeed it may very well have been just that, with its travel restrictions, fear, political mobilizations (co-opting a virus for social ends and shifts of power), economic disruptions and closures, social stress, declaration of a

world health emergency, and tamping of contrary narratives reflective of doubt.

The success of Brazil's long-planned, heavily-invested, and joyfully-anticipated 2016 Rio Olympics was jeopardized and downgraded. Predominantly-Catholic Brazil's abortion-prohibitions were circumvented and called into question.[20] The embattled (and impeached) president (Dilma Rousseff) tried to rally her base by calling out the military, and applying substantial public health measures and restrictions.[21]/[22]

Outside of Brazil, fears persist: the potential ruination of a future-child's life; with resultant severe mental retardation from merely a chance insect-bite's strange-sounding, unknown virus. Yet, for all the doomsaying, the predictions surrounding Zika never came true. Despite this, "science" has never retracted the Zika-microcephaly theory. There have been some contortions of announcing variant strains or underlying immunity as having forestalled any recurrences in Brazil, or onset anywhere else. Let's investigate what happened during the Zika pandemic and discover what we can learn from it, today.

The results may surprise.

From: "The Zika Virus Epidemic in Brazil: From Discovery to Future Implications"

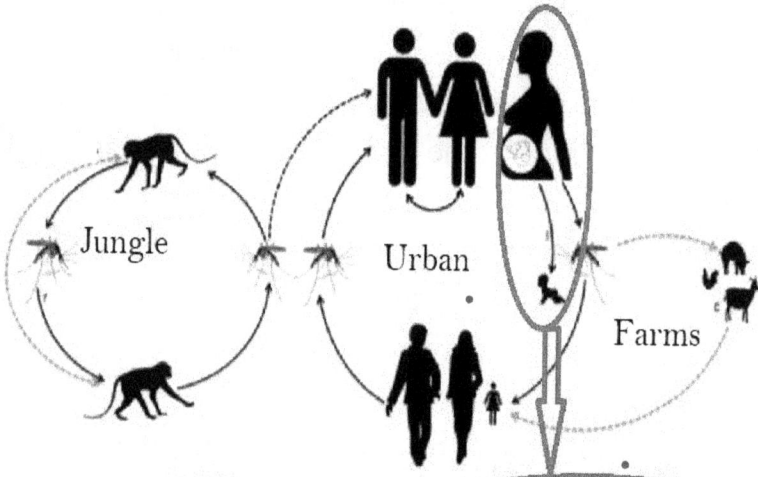

Jungle

Urban

Farms

3%-8% attack rate?

Zika-
Microcephaly?

Microcephaly is not always a precise diagnosis. There were multiple different measurement standards for determining microcephaly within Brazil at that time.

Normal head size

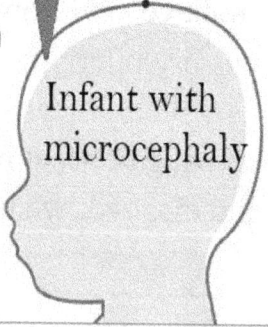

Infant with microcephaly

Source: ADAM, WHO

Zika, A Shock to The System

Unlike COVID-19, which is a rebirth (of sorts) of a KNOWN illness (2003's original SARS coronavirus) -- Zika's claimed potential to cause congenital microcephaly was *"a shock to the system"* from a virus essentially completely unknown socially and even medically (except to tropical-disease specialists). For the entirety of its known history, dating from its 1947-identification in Uganda, Zika had caused a grand total of ZERO prior, grave problems in humans.[23] [24] [25] West African populations, tested for Zika as early as the 1970s, had some background immunity, with no records of increased microcephalic births to infected mothers.

So, when the previously completely benign Zika virus was believed to have taken on this malignant role in pregnancies, it was understandable that warnings and advisories should be made and broadcast. In retrospect, there should have been a greater emphasis on comparative data analyses and validated scientific review -- along with overall skepticism, absent either of these. It was (and remains) highly improbable that Zika — (alone among its Flaviviridae family[262728]) causes birth defects; moreover, so very suddenly after 60 years of entirely innocuous human contact.

Flaviviruses are by no means rare. The most "famous" of the flaviviridae family, Hepatitis C (HCV), infects millions around the globe -- without any microcephaly; in fact, without birth defects at all:

> *"Overall, there is no apparent deleterious effect of pregnancy on the course of HCV infection: no evidence to suggest an increased number of congenital anomalies in children born to HCV-infected women."*[29][30]

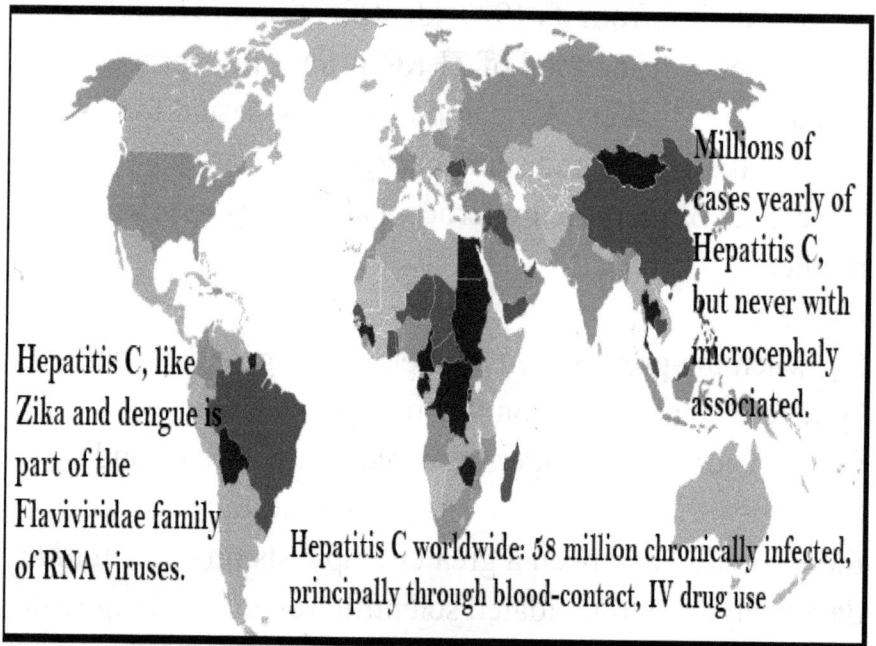

Millions of cases yearly of Hepatitis C, but never with microcephaly associated.

Hepatitis C, like Zika and dengue is part of the Flaviviridae family of RNA viruses.

Hepatitis C worldwide: 58 million chronically infected, principally through blood-contact, IV drug use

The arbovirus ("*ar*thropod-*bo*rne *virus*", i.e. mosquito-transmitted) flavivirus subset: Yellow Fever, Dengue, Japanese encephalitis, and West Nile --

> *"have been associated with human disease since ancient times. Reliable historical descriptions of epidemics caused by what appears to be the yellow*

fever virus have been documented as early as the mid-1600s."[31].

Their dangers have, through the centuries, never been endured by the recipient's gestating embryos (protected by the placental barrier).[32] The CDC states,

"*Flavivirus infection during pregnancy has not been known to cause birth defects in humans*."[33]

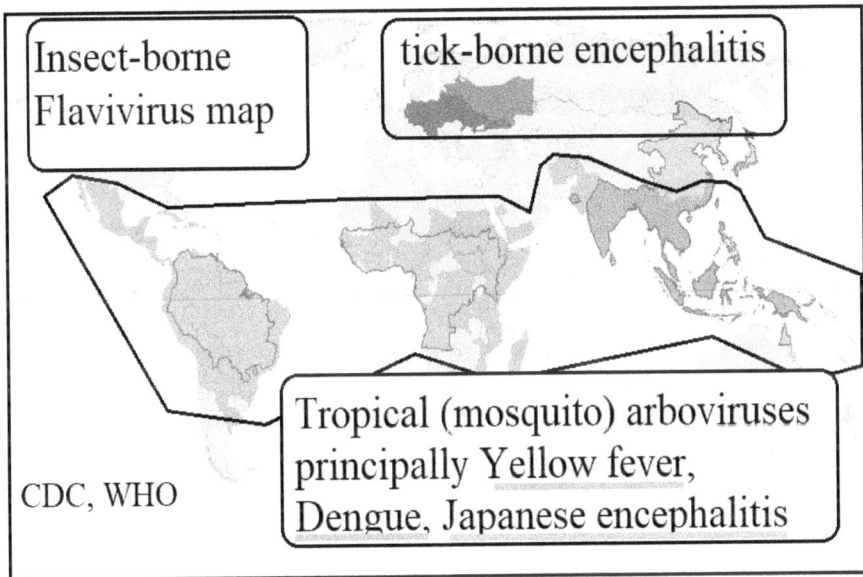

Insect-borne Flavivirus map

tick-borne encephalitis

Tropical (mosquito) arboviruses principally Yellow fever, Dengue, Japanese encephalitis

CDC, WHO

Zika remained unknown -- because of no pressing urgency *to* know it. At the time its health ministers declared a public health emergency, Brazil (like essentially every other country on earth) had had ZERO previous record of Zika not only as an illness-- but *AT ALL*! Even during the pandemic year, 2015, *there was* **NEVER** *a single active Zika patient diagnosed*!

In summary, Zika had never been seen, tested, recorded, or diagnosed as a human disease in the Americas (North or South) up to and INCLUDING its presumed pandemic year, 2015. No virus in its flavivirus (family or genus) has ever been associated with birth defects generally -- let alone specifically this one: microcephaly. It was therefore medically improbable for Zika to be anointed as a microcephaly-cause. A separate improbability covered later is that this Brazilian microcephaly increase hadn't been statistically documented, at the time it was announced.

Zika, dengue: *"separated at birth"*?

In the lead-up to the 2015 Zika-microcephaly pandemic in Brazil, there were minimal rumblings in the academic medical community of first-time-ever Zika illness after a 2007 episode in the Yap Islands (east of the Philippines), where there was an illness characterized by rash, conjunctivitis, and arthralgia (joint pain, the hallmark of dengue fever, a.k.a. "bone break fever[34]"). In fact, Yap's physicians received laboratory confirmation of their clinical suspicions of ***dengue***. Months later, however, the CDC's Arbovirus Lab's RT-PCR testing suggested a match with Zika.

When referees call a game, close plays may get reviewed, in which case "the ruling on the field" may be overturned. Something similar occurred when the CDC overruled Yap physicians' on-site clinical and laboratory diagnosis of dengue. The CDC's laboratory result announcement didn't consider symptoms - or weigh in on Zika's long-standing harmlessness as opposed to dengue's persistent danger; let alone the confusing and confounding laboratory cross-reactivity between the two.

Zika and dengue are nearly indistinguishable, visually, genomically and molecularly.[35] The Rockefeller University's Science Outreach program helps visualize this Dengue-Zika similarity with creative "Virus Origami"[36] (using models from the Research Collaboratory for Structural Bioinformatics (RCSB) Protein Data Bank (PDB)), capitalizing on Flaviviruses' icosahedral symmetry.[37] For those unfamiliar with this shape, a soccer ball is a "truncated icosahedron", with its resulting black pentagons' reflecting each five-triangle junction.[38]

Flavivirus "origami"

Build a Paper Model of Zika Virus

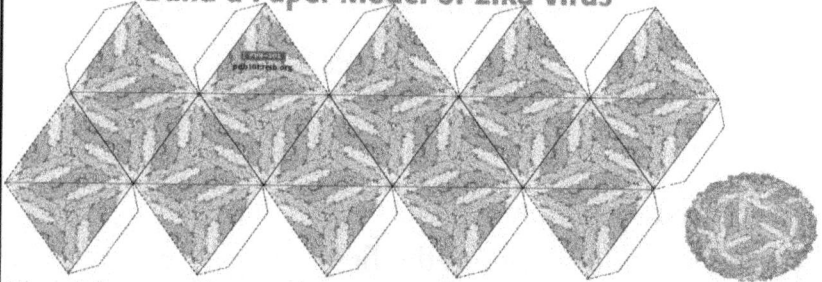

Flaviviruses have icosahedral symmetry (a bit like a soccer ball's). there is a 20-fold repetition of a single triangular pattern: Dengue and Zika, both.

Build a Paper Model of Dengue Virus

The actual proteins show 50% identical overlap between Dengue/Zika

Put the "origami" shapes together and you get a reasonable approximation of the dengue and Zika virus protein shells.[39]

Showing the three- & five-part symmetry

In both dengue and Zika

from"... ZIKA versus Dengue Virus Protein Shells" Sci Rep 10, 8411 (2020)

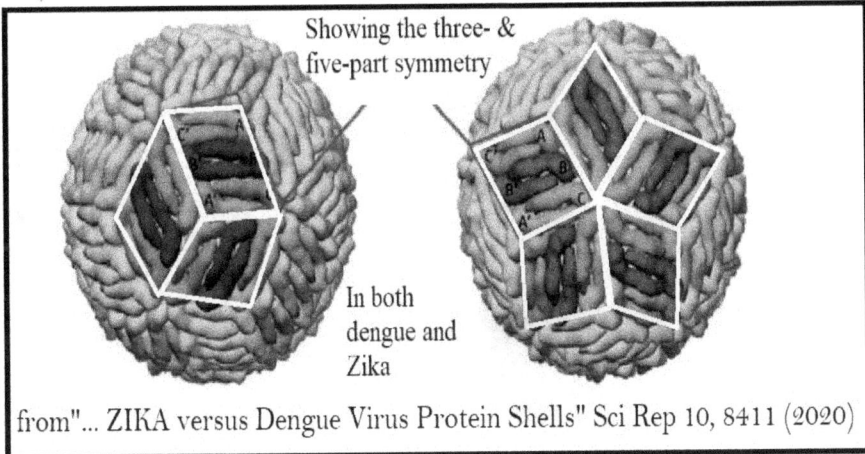

This Zika-dengue similarity has recently been reinforced in studies showing that immunity to one may cover the other. In a 2020 study showing the "impact of flavivirus vaccine-induced immunity on primary Zika virus antibody response in humans", [40]there was observed a "strong anamnestic response in pre-vaccinated individuals". That means there was an enhanced reaction of the body's immune system to Zika antigens because they were related to an Flavivirus or dengue antigens previously encountered. Here is a diagram from that article showing the very close relation between Zika and dengue structural proteins.

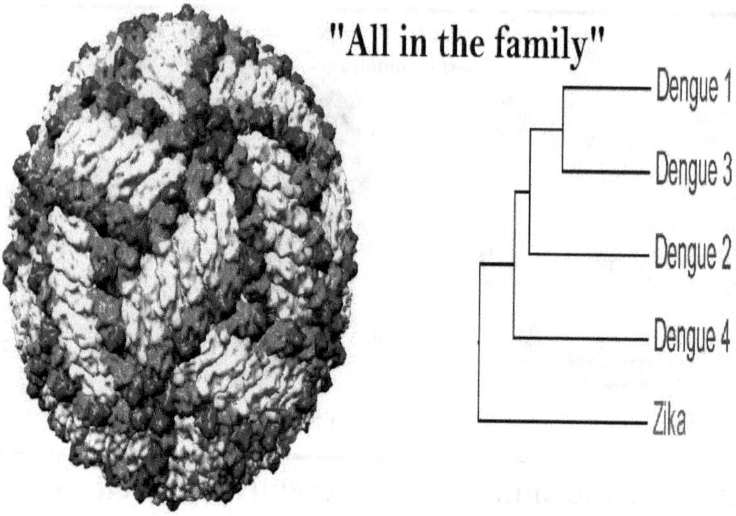

"All in the family"

- Dengue 1
- Dengue 3
- Dengue 2
- Dengue 4
- Zika

"Many people at Zika-risk are pre-immune due to prior infection with dengue virus (or other flavivirus vaccinations)."

from: Impact of flavivirus vaccine-induced immunity on primary Zika virus antibody response in humans, *PLoS Negl Trop Dis 14(2) 2020*

Here's a graphical representation of the comparable protein domains between differing strains of Zika and dengue[41]. The closer the lines become, the greater is the similarity.

Comparison of conservation of flavivirus E-protein structure and sequence between Zika and dengue. (I.e. very similar)

from Structural biology of Zika virus and other flaviviruses
Nature Structural & Molecular Biology | VOL 25 | JANUARY 2018
| 13–20 | www.nature.com/nsmb

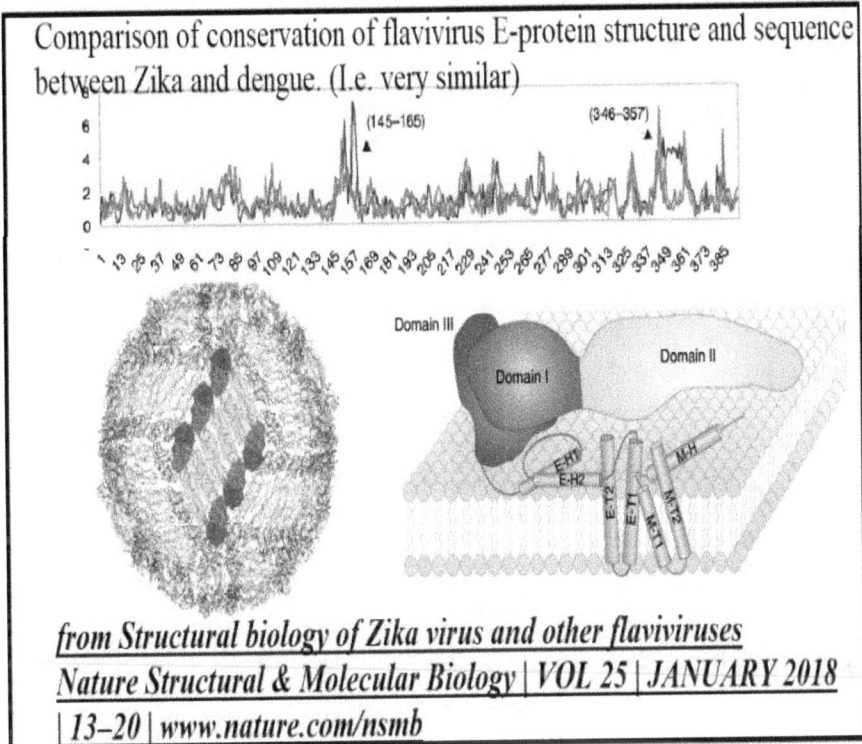

Zooming in even more closely, a recent (2021) study showed an ordered lipid that is present in only the mature forms of Zika, dengue (and its less prevalent Flavivirus cousin Spondweni virus). [42] The similarities down to the molecular level probably make Zika and dengue closer than any two chocolate chip cookies from the same batch. Each spiral is a protein, essentially identical in one virus to the other – but not seen in other viruses outside this trio.

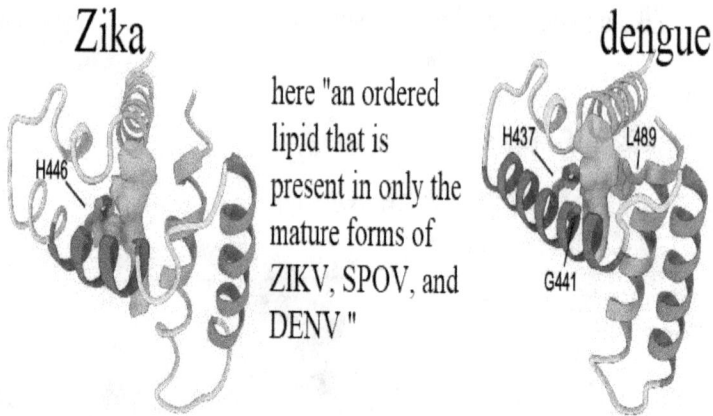

Zika-dengue similarities on a molecular basis

Zika   dengue

here "an ordered lipid that is present in only the mature forms of ZIKV, SPOV, and DENV "

H446   H437   L489   G441

From: Flavivirus maturation leads to the formation of an occupied lipid pocket in the surface glycoproteins
NATURE COMMUNICATIONS (2021) 12:1238 https://doi.org/10.1038/s41467-021-21505-9

Amongst these biologic twins, mistaken cross-diagnosis is nearly a given. This dengue-Zika cross-reactivity had never been considered an issue because only dengue was active and tested. It was not until the year AFTER Brazil's Zika-microcephaly panic that a certain few clinical laboratories even began to stock and apply Zika-tests.

Nonetheless, the medical academy went with the influential CDC's call, activating Zika from a previously well-earned dormancy; reassigning it a new role as viral agent of mystery. Without the other flaviviruses' centuries of clinical familiarity, Zika was available as an empty vessel - - to be filled with unproven powers its well-known viral cousins lacked.

From this hasty reckoning's elevating Zika into a disease entity, any of three scenarios are possible:

- Zika and dengue are not just close genomically, but clinically as well: implying that before advanced testing existed, both were included under the rubric of "dengue", a situation the CDC (via its Yap ruling), rectified. *OR*...

- There had been an urge to cut back on dengue's reach, incorrectly attributing genuine dengue cases to a more exotic (and research-fundable) novelty, Zika. *OR*...

- The CDC laboratory was in error, cross-reactively showing Zika when in fact it was dengue.

It's a puzzle possibly with no current answer. It may also be a *"distinction without a difference"*[43] insofar as both are flaviviruses transmitted by the Aedes mosquito.[44]

Despite news of Zika's first ever (presumed) human disease causation in 2007, it remained (for the meanwhile) far from people's consciousness. Throughout the early 2000's, Google Trends showed only the slightest bulge of a handful of Zika-searches, worldwide. To put this in context, the obscure term *"Yap"* itself was ~100 times more popular, but not for long. In Zika's "comet phase" (at its height of fear and notoriety), it was at near parity with *"Brexit"*, another "out of nowhere" term from 2016.

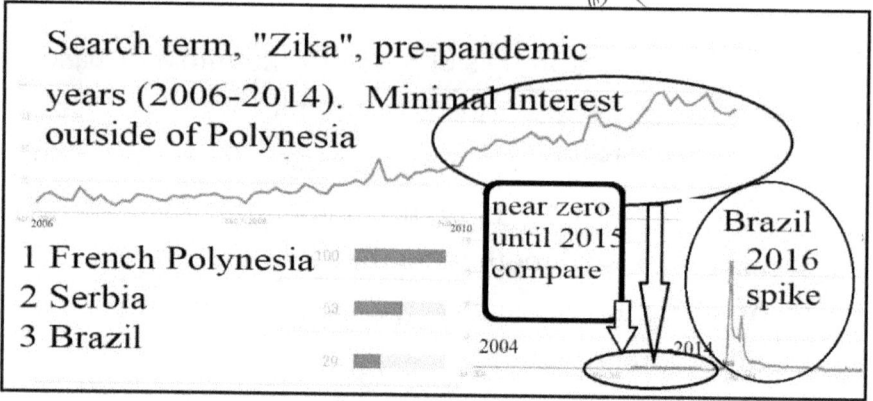

Search term, "Zika", pre-pandemic years (2006-2014). Minimal Interest outside of Polynesia

near zero until 2015 compare

Brazil 2016 spike

1 French Polynesia
2 Serbia
3 Brazil

Between its 1947 discovery in the Zika Forest, Uganda and the early 2000's, there were a few "sightings" immunologically, but not clinically.[45]

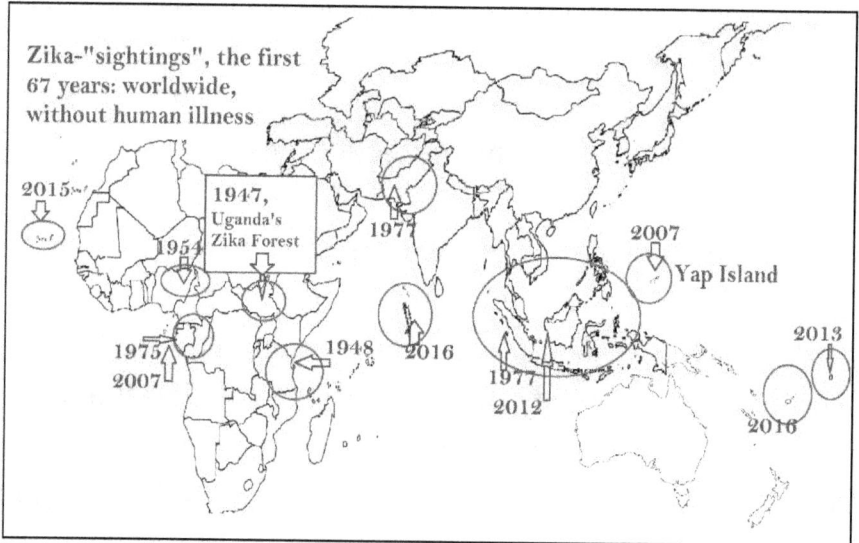

Zika-"sightings", the first 67 years: worldwide, without human illness

1947, Uganda's Zika Forest

Yap Island

Notably, the map shows distinctly different geographies' underpinning the two Zika subgroups: the "Africa-strain" and the "Asia-strain". There's no "Americas-strain" to be

found-- and no history of Zika in the Western Hemisphere, prior to Brazil, 2015.

There was, however, dengue – endemic in Brazil since its transfer from Africa some decades or centuries prior. Dengue's necessary first ingredient, the mosquito vector, *Aedes aegypti* may have arrived as early as Brazil's 16th century African slave trade, alone or in tandem with its arbovirus, dengue, which wasn't serologically noted until the 1980s.[46]

Brazil in 2015 without broad Zika-testing capabilities or an internal supply of Zika-primers, had no formal algorithm with which to distinguish Zika from dengue. It wasn't until 2019 that this particular *"Dengue vs. Zika"* algorithm was developed. [47]

Its intricacy points out the frivolity of making a differentiation claim of Zika (versus dengue) -- not only without testing, as was the case in spring 2015 – but even with testing! This algorithm underlines the need for top-quality reagents and technicians expert in this practice, a situation which could not have existed in Brazil until a few years later. Even years after the Zika "pandemic", there is continuing research specifically to develop biomarkers' differentiating Zika from dengue.[48]

No easy task to differentiate Zika from dengue

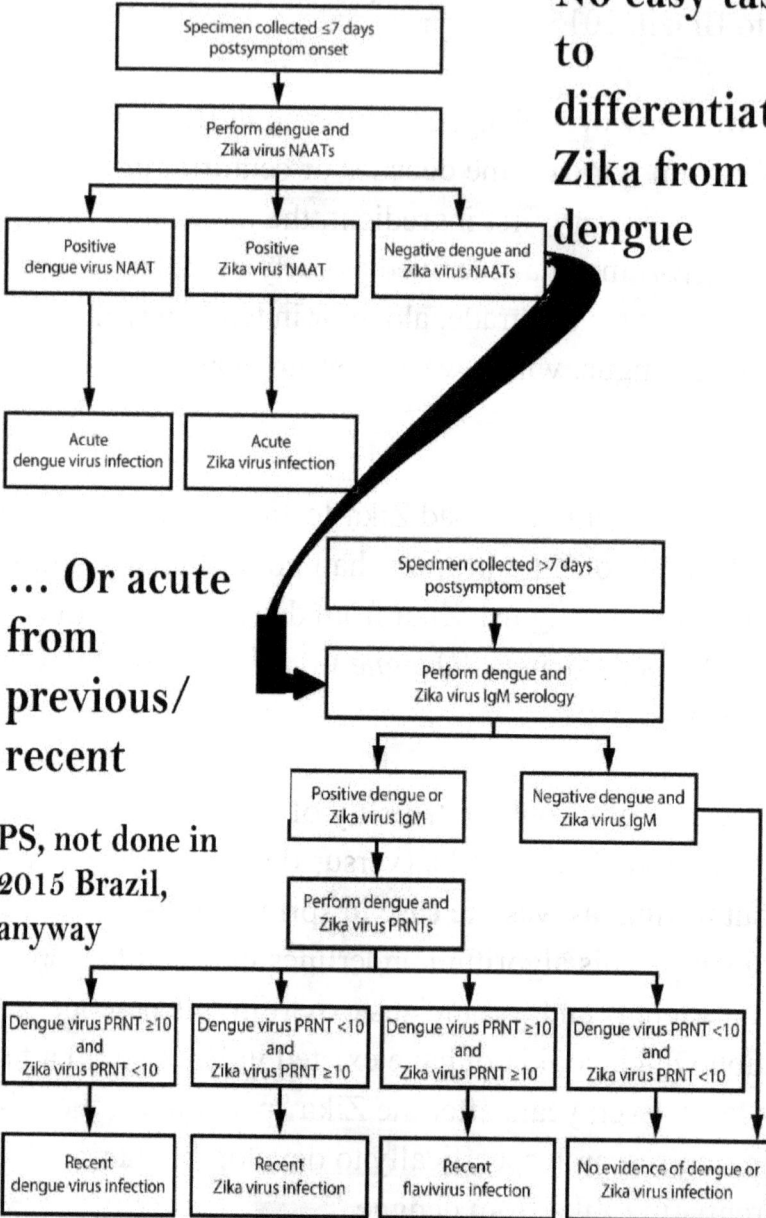

Specimen collected ≤7 days postsymptom onset

↓

Perform dengue and Zika virus NAATs

↓

- Positive dengue virus NAAT → Acute dengue virus infection
- Positive Zika virus NAAT → Acute Zika virus infection
- Negative dengue and Zika virus NAATs

... Or acute from previous/ recent

PS, not done in 2015 Brazil, anyway

Specimen collected >7 days postsymptom onset

↓

Perform dengue and Zika virus IgM serology

↓

- Positive dengue or Zika virus IgM
- Negative dengue and Zika virus IgM

Perform dengue and Zika virus PRNTs

↓

Dengue virus PRNT ≥10 and Zika virus PRNT <10	Dengue virus PRNT <10 and Zika virus PRNT ≥10	Dengue virus PRNT ≥10 and Zika virus PRNT ≥10	Dengue virus PRNT <10 and Zika virus PRNT <10
Recent dengue virus infection	Recent Zika virus infection	Recent flavivirus infection	No evidence of dengue or Zika virus infection

Dengue and Zika Virus Diagnostic Testing for Patients with a Clinically Compatible Illness and Risk for Infection with Both Viruses, 2019 MMWR

"Medicine is a social science, ...

*...and politics is nothing else but medicine on a large scale." **Rudolf Virchow, 1848***

Dr. Virchow as the 19th century's "Pope of medicine" and later politician felt that "*physicians are the natural attorneys of the poor, and social problems should largely be solved by them.*"[49] [50] Likely 2015-Brazil's central actors -- the physicians and scientists developing the triad of

- Zika-as-illness,
- observation of a microcephaly-increase; and then
- a Zika-microcephaly causation linkage--

align with Dr. Virchow's thoughts of social activism through medicine; leaving aside the danger that, without firm foundational science, medicine may become a house built on sand. Inviting political thought as a guide for determining medical goals runs the separate risks of overreach and manipulation.

Arguably this has occurred at places and times within the more recent pandemic. The best "science" (in both senses of the term): provable facts; as well as the scientific method necessary for proof – can only come about in a lasting sense, if its creation is not too heavily reliant on social, personal, or political leverage on the observation and recording of facts. Science isn't poetry or politics– and although medicine can be used socially, the science underlying it can't be wished into existence through popular will.

Albert Einstein was asked

"Why is man (able to discover the atom's force) unable to devise the political means to keep it from destroying us?".

Einstein answered:

"Simple, ... politics is more difficult than physics."

What would Einstein say about the Zika-microcephaly "creation story"? Probably he wouldn't find the science so much "difficult" as ignored or manipulated. On some level it's fairly straightforward: *a virus, an infection, a side effect*– but the detection of, and an agreement on, the facts become rather murkier. **Overturning Zika** hopes Mr. Einstein would have appreciated this "simplification" diagram. It might be a little bit too much to digest all at once at this point, but the information will come later in **Overturning Zika.**

Zika-microcephaly, 2015, Northeast Brazil

Mosquito → early pregnancy

Zika → birth

Literally zero Zika tests performed clinically in Brazil 2015

(Or dengue?)

no prior microcephaly registry or uniform standards / metrics

Millions of cases of dengue had never caused any microcephaly

Microcephaly found!

but later, with firmer standards

Previous years microcephaly rate

2015's microcephaly found not much different from...

"Upon further review", no real change (2015 or 2016) from prior years' microcephaly rates, but no retractions

Microbiology and medical diagnoses are not quite as straightforward as particle physics or astronomy. Einstein's theory of relativity may at first seem like magic or confusion to most people – but it is very clearly definable and provable, once the facts and premises are aligned for comprehension.

Zika-microcephaly perhaps has the opposite effect, it seems less and less clear the closer one looks. Echoing our earlier point, dengue and Zika virions are essentially indistinguishable one from the other:[51]

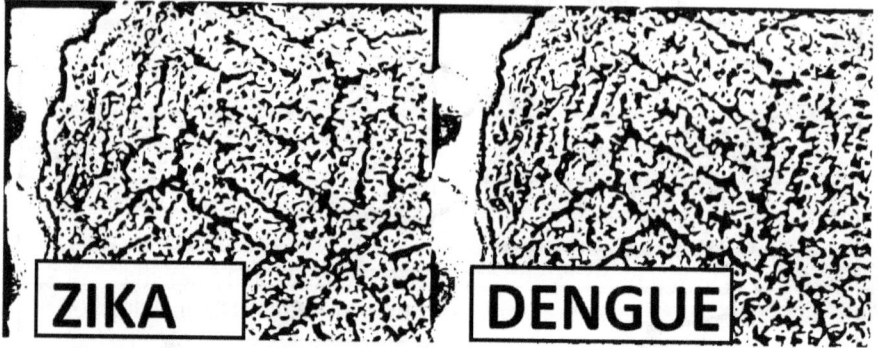

Absent Brazil's pre-existing widespread dengue disease, Zika-as-illness may never have been locally conjured. Perhaps inspired by the 2007 Yap example (of clinically diagnosed dengue's being relabeled as Zika), pediatrician Dr. Kleber Luz[52] focused on milder cases of dengue in Brazil, early 2015, *"and **insisted** that it was something urgent and new (Zika)"*[53] despite having a grand total of zero confirmatory lab tests to warrant such confidence. Demonstrating that medicine's not only partially a social science but also a social grouping, he was readily joined by others with a similar zeal and mission to find a clinical role for previously ignored Zika.

Prior to those thoughts of Zika, Kleber Luz had joined up with Dr. Carlos Brito in October 2014 to form a WhatsApp group *"CHIKV, The Mission"*,[54] to discern (the anticipated) presence of a brand-new virus within Brazil, "chikungunya" (**CHIKV**), which may have been on the march globally from its native Africa, without having yet

reached Brazil. [55] Here are some brief bios of the five central medical actors in the Zika-as-illness story, Brazil, 2015:

- **Silvia Sardi & Gübio Soares Campos (S&SC):** Virologists joined both in marriage and professionally as professors at the Federal University of Bahia. In April 2015, the two ran lab tests with off-label primers that claimed Zika as both a new disease entity, and new Brazilian arrival.
- **Claudia Duarte dos Santos:** Virologist at the Oswaldo Cruz Institute (Fiocruz) in southern Brazil, in contact with Dr. Kleber Luz.
- **Kleber Luz:** Pediatrician and epidemiologist from Natal, Rio Grande do Norte, who in 2014 hypothesized chikungunya was newly circulating in Brazil, forming a loose task force CHIKV, The Mission – before changing focus to potential Zika.
- **Carlos Brito**: General practitioner and epidemiologist from Recife, Pernambuco, who collaborated with Dr. Kleber's hunt for chikungunya, then Zika. When, completely separately, it was surmised that there was more microcephaly in Recife, this local physician was called to the scene. Dr. Brito investigated and concluded a connection between his two projects: Zika and microcephaly.

These physicians took their group's name and inspiration from the 1986 film, *The Mission* – to capture the zeal, fervor, devotion and direction of the story's missionaries in 18th century Paraguay venturing into hostile indigenous tribes' territory. **"CHIKV, The Mission"** added interested infectious disease-, epidemiology-, virology-, and clinical professionals in Northeast Brazil; those somewhat off the main, institution-based academic grid. The group may not have been entirely selfless in its hunt for some new disease entity, the discovery which would raise the profiles of its members. It's doubtful the general public or physicians (in general) clamor for any new diagnosis, absent a cure. Just relabeling certain milder cases of dengue as chikungunya (or a much-later realized Zika) achieves more in changing the status of the diagnosticians than the patients.

Within a few weeks of Dr. Kleber Luz' (either intuitive or groundless) insistence of dengue cases' being Zika (early spring 2015)– completely separately, a husband-and-wife team of Bahia virologists, Drs. Silvia Sardi and Gubio Soares Campos (**S&SC**), made a similar "Zika" - declaration. Had they gotten wind of the other doctors' claims through WhatsApp's **"CHIKV, The Mission"** or some other conduit?

In the shadow of Brazil's largest petrochemical plant[56], **S&SC** chose to probe for Zika as an explanation of Camaçari, Brazil's ~5000-patient set of mildly ill "allergy" patients. Their laboratory "confirmation" was of the

slenderest variety: seven unconfirmed positives from a pool of 24 samples unassociated with clinical severity. And even those "positives" were shaky. Their small laboratory's Zika- reagents were off-label, unconfirmed and uncertified. No surprise then that only half of even that small number was backed up by later independent testing.

Nowhere (else) in Brazil but with **S&SC** did there exist in early 2015 Zika PCR- "primers" on hand, at the ready. Serendipitously(!) and mysteriously, the virologists' set had fallen into their laps, left by a visiting Senegalese researcher the year prior[5758] never to be confirmed by Brazil's "FDA" or other researchers for efficacy or accuracy.

In a pattern that was duplicated in the "microcephaly" -half of the Zika-microcephaly saga six months later, a medical revolution (*"Zika's first-ever human illness!"*) came about through sensationalism rather than science. **S&SC** leaked their suppositions to the press[59] before any other laboratory could analyze or question the results (using higher quality reagents and primers). This showmanship, leaving no opportunity for critical review, falls far outside normal scientific protocols. As defense, Dr. Soares Campos reports a green light given by certain Ministry of Health (**MoH**) officials:

> *"We decided to **benefit the public** more, rather than immediately writing a scientific paper and publishing it."* [60]

It is difficult to see precisely what benefit Dr. Soares Campos was providing insofar as in his own view at the time he saw, *"Zika Virus as not as serious as dengue or chikungunya, it does not lead the patient to death. The condition seems allergic, it is calmer and the treatment is the same. The treatment is the same for dengue: Paracetamol. You don't fight the virus. Your body does it. You fight the symptoms."*[61]

Here is the original article, leaked to the press April 29, 2015:

≡ MENU   **G1**   BAHIA   REDE BAHIA

April 29, 2015

29/04/2015 17:32 - Updated 29/04/2015 17:32

Virus causing a mysterious disease identified in Salvador and RMS

Symptoms are similar to those of dengue, but less severe.
Researchers believe that the virus arrived in Brazil during the World Cup.

Two researchers from the Institute of Biology of the Federal University of Bahia (UFBA) discovered the virus that causes the disease, whose symptoms are similar to those of dengue and which has been frightening the Bahian population: the Zika Virus, which is transmitted by the mosquitoes Aedes aegypti, Aedes albopictus and other types of aedes.

According to Gúbio Soares, a researcher who made the discovery together with Silvia Sardi, it is the first time that the virus has been identified in Latin America, being more common in Africa and Asia.

This article mentions additionally that *"the discovery by Gúbio (Soares Campos) and Silvia overturns the two hypotheses raised by the Epidemiological Surveillance and the Camaçari Health Department to explain the disease.*

Last March, the two bodies suspected that the symptoms would be caused by roseola or parvovirus-B19." In any event, nobody thought Zika was dangerous at the time, thus there was no urgency.

This is a workable approach to public relations, but not so much to science. Dr. Soares Campos should have had a stronger allegiance to the scientific method and standards of proof. He and his wife were the PhD-professors on that call. That's not to say they necessarily were wrong in their Zika-findings, but if there was true confidence in the veracity, the real *"benefit to the public"* would accrue with and after outside research confirmation.

Mind you, there was no hurry at this point, unless it merely was to beat out the **CHIKV** group, hot on the trail of finding something to call Zika. The Camaçari disease entity they relabeled was (if viral) at worst mild dengue or some measles variant from which people recovered spontaneously. And if the problem had emanated from a petrochemical toxin, then their work was a distraction. Their announcement brought no cure or change in treatment and fell also into the category of a medical "distinction without a difference". The researchers and bureaucrats may have benefited, but in what way did the public?

"Haste makes waste" vs. "Too little too late"?

Within a month of the unverified but public announcement of Zika-as-illness by Bahia virologists Drs. Sardi's and Soares Campos (**S&SC**), Bahia-State's health ministry (**SESAB**) publicly contradicted them with a statement that the ***"diagnosis of Zika cases in Bahia may be wrong."***[62] **SESAB** made a very compelling case:

- **S&SC's** (Drs. Sardi's and Soares Campos') primers had been illegally imported without authorization or review.
- **S&SC**'s flaws in handling primers could have led to mistaking Zika for actual dengue or false positives.
- ***More than half of a separate, larger collection of 500 blood samples taken from patients diagnosed with the "mystery disease" tested positive for dengue.***
- None of that much larger group of dengue-positives reacted to Zika.
- A larger number than Zika-positives (10% of the more than 5,000 cases of mild fever and rash) had tested positive for rubella, measles and Parvovirus.

The speed with which **SESAB** elaborated these weaknesses in **S&SC's** argumentation points out the original flaw in the virologists **S&SC's** releasing information directly to the public without scientific review. Clearly, **S&SC** could have waited. The world wasn't on edge over whether "mild dengue" was actually "Zika". **S&SC** would have been forced to deal with and explain away **SESAB's** objections and potentially others' -- if they had gone through proper channels, submitted their findings to peer review, and maintained patience. The trade-off would have been a month or two delay (if they were right) versus not feeding the public false information (if they were wrong).

SESAB's first two points address the virologists' haphazard and unsystematic methodology of using unverified primers passed down by hand. Nearly half of **S&SC's** small number of Zika-positives were overturned, leaving only 12% of the 24 Camaçari patients' blood samples' showing Zika (that is, only four individuals in a city of 300,000 people) -- and any or all of those four may have been either dengue misdiagnosed, or nothing at all.

Let's look at **SESAB**'s crucial third point. Looking at the same Camaçari illness, SESAB found **250 positive dengue tests** to the **S&SC** virologists' **4 Zika positives**– at a rate of **50% to 12%**. The fourth point shows the dengue positives' excluding Zika. Which, then, of the two, (dengue or Zika) would appear to have the stronger claim as the underlying cause of what in fact physicians had originally conceived as "mild dengue"?

SESAB's fifth point underlines that the **S&SC** virologists' presumed Zika didn't show up at any greater rate than other illnesses with longer histories of producing mild fever and rash: rubella; measles; parvovirus. Yet, what's the narrative that all of Brazil remembers, coming out of this incident? Of course, it is the ***new mysterious virus, Zika!*** It's hard to highlight a better example of social pressure and consensus' determining a new medical scientific "fact".

Those cautions didn't factor into the jubilation amongst the researchers nor the acceptance by a less scientifically cognizant public excited by the press' eternally emphasizing the "new" in "news". Drs. Sardi and Soares Campos outlasted the criticisms and became "discoverers of Zika" (*as an illness*)[63], parlaying this bold move into tenure positions and medical prominence.

Instagram   gubio.soarescampos

352 posts   1,590 followers   451 following

Gúbio Soares Campos
Scientist
 Doutor em Virologia pela Universidad de Buenos Aires
 Descobridor do Zika vírus no Brasil
www.facebook.com/gubiosoaresc

Dr. Kleber Luz lost the race (almost literally for fame and fortune), but not for lack of trying. He had implored two major institutions to produce results matching his wishes (for Zika).

> *"Unhappy with the Evandro Chagas Institute's inconclusive findings, Dr. Kleber Luz contacted the*

Oswaldo Cruz Institute in Paraná, southern Brazil, where Dr. Claudia Nunes Duarte Dos Santos received the blood samples with a mandate: "It's Zika. Find Zika," Dr. Kleber Luz had told her." [64]

Dr. Dos Santos deserves credit for not compromising her craft: finding no Zika despite an order to do so -- although she later wrote:

"I am so very sorry we failed. But none of the samples you sent came out positive. ... You were really very skilled. If you ask me who discovered the Zika virus in Brazil, I'll tell you: 'Kleber Luz.'"[65]

Dr. Kleber Luz apparently felt the same way because by June he had published a misleadingly titled Zika-discovery article of his own co-authored by his laboratory confidante.

First report of autochthonous transmission of Zika virus in Brazil

Camila Zanluca , Vanessa Campos Andrade de Melo , Ana Luiza Pamplona Mosimann , Glauco Igor Viana dos Santos , Claudia Nunes Duarte dos Santos , Kleber Luz

"In early 2015, several cases of patients presenting symptoms of mild fever, rash, conjunctivitis and arthralgia were reported in northeastern Brazil. ...**This is the first report of ZIKV infection in Brazil.**"

Mem Inst Oswaldo Cruz, Rio de Janeiro, June 2015

Dr. Kleber Luz gives no credit in his article to the Bahia virologists. How this squares with Dr. Dos Santos' apology for not conjuring Zika from blood samples a couple of months prior is hard to say.

Nonetheless, between these two pronouncements, the Zika-as-disease narrative was taking hold, shortly echoed by Recife's Dr. Carlos Brito[66] who similarly inferred *"[that 80% of Pernambuco's dengue cases are in reality] Zika. We investigated 1,100 (dengue) patients.* **Of this total, 81% met clinical criteria for Zika.** *"*[67]

Dr. Brito (it appears) was characterizing certain (milder) cases of dengue as Zika,
- without having done any confirmatory testing;
- without having examined the patients
- without Brazil's having had a long enough history of Zika to determine "clinical criteria", even if you believe the Bahia virologists' highly doubtful version
- without Zika's having developed generally accepted "clinical criteria" anywhere else (except low-likelihood Yap, 2007)

Understandably, there were some behind-the-scenes doubts, in government and the academy.

> *Drs. Brito and Luz contacted Brazil's Ministry of Health, but the Brazilian government wasn't*

*convinced. Whatever was going around had no dangerous symptoms or long-lasting effects, and people weren't dying. So the government chose not to implement mandatory reporting for the infection, and when the summer of 2015 ended, so did any worries about Zika. "There was enormous resistance to the idea that it could be Zika." **Frustrated, the doctors** (Brito and Luz) **decided to form their own group to study the virus.**[68]*

It's practically human nature to root for the underdog; however, it could be that here academic and institutional resistance, doubt, and caution were the better recent propositions. Nonetheless, there's no stopping a good story: Zika, like Frankenstein's monster -- was stirring into life, eventually to bring similar fears.

The Eureka Moment, Microcephaly in Recife

My own central medical experience emanates from 30
years as a primary-care physician in the United States. That
experience comes with a disclaimer in regard to evaluating
the Zika-microcephaly story: throughout that time I have
never treated Zika-exposed mothers or microcephalic
infants – nor have I ever traveled to Brazil. I have had no
association with Zika (or any other) medical research.

So why, then, did I latch onto -- and attempt to investigate,
and potentially solve -- a problem that nobody else really
seems even to think of as a mystery in need of solving? We
may have to ask my wife, sons, friends or colleagues, but
(for me) it comes down to two things: respect for the
scientific process, and enjoyment in solving puzzles. Truth
be told, generally I have enough puzzles and problems in
my path at any given moment that I don't look for more.
But somehow, this was different.

In 2016, I had kept up with Brazil's Zika-microcephaly
news. Both aspects of the connection seemed puzzling.
Zika -- previously medically unknown and harmless;
absent from the Americas, but a specific single entity-- was
connected to microcephaly, a condition in almost every
way its opposite. The only thing they did have in common
was extreme rarity. Microcephaly (like other congenital
abnormalities) is a "fact of life" for all vertebrates, a rare

but timeless abnormality noted in humans (but not zoologically because such animal offspring cannot survive in the wild). Aside from one rare, specific genetic syndrome, severe "true microcephaly"-- microcephaly has no single specific cause, only the potentiality of a vast number of minor, general, toxic influencers; none of which usually can be diagnosed after the fact.

Even so, I would have remained a bystander as the story played out but for a quirk in the origin of the Zika-microcephaly theory. Dr. Vanessa Van der Linden Mota, the Brazilian neuropediatrician based in Recife credited with connecting Zika to microcephaly claimed her "*Eureka moment*"[69] [70] of August 2015 occurred examining fraternal twins, one with microcephaly, the other healthy.[71] "*Testing (for congenital defects) turned out normal, the doctor said,* '*an indication that we were dealing with something new.*'"

Her "*Eureka moment*" essentially brought about my own, but in the opposite direction. Her conclusion—that simultaneous, coexisting normality and microcephaly in fraternal twins (with their separate eggs, separate genomes, and separate placentas) necessitated a completely new disease cause —*itself* seemed problematic. It is actually rather an extraordinary leap simultaneously to

- ignore the hugely anecdotal nature of the circumstance from which to draw such a bold

inference (including ignoring the twins' doubled-placenta itself potentially being a cause)

- attribute maximal danger to a previously benign virus, never documented even once to that point as even existing in the Americas[72]
- assign it a potential to cause congenital microcephaly not shared by any of its Flavivirus relatives, including dengue (Zika's own fraternal twin) which had been infecting Brazilians in the millions, per year.[73]
- do so without any host tissue or virologic testing confirmation
- avoid considering the contrary evidence of the normalcy of the (presumably similarly infected) other twin
- AND assume a false certainty that the unfortunate child's situation couldn't have been caused by any of microcephaly's dozens of other not-precisely-diagnosable causes.

That was what was going through my head. Since researching this book, I have found that her story (like almost every other) is more involved and complicated, and that her "Eureka Moment" (believing one twin had some type of infection) was aggrandized into the bold declaration of a microcephaly epidemic by many circumstances out of her control. I also learned that she had had some cautions, and that the "final touch" of assuming that Zika was the root cause was applied by Dr. Carlos Brito, not Dr. Vanessa van der Linden.

The Family Business

Neuropediatrics has been the "family business" for at least two generations of Recife's *van der Linden* family: practiced locally by Dr. Vanessa van der Linden and her mother Ana and in central Brazil by her brother, Helio. The severely microcephalic twin that started her thoughts of a new causation for microcephaly, her so-called "patient zero" was noted to have excess scalp skin, and certain skull calcifications of a type she had come to associate with positive infection-panels in previous cases.

[*The microcephaly events and timeline described in this chapter were gathered from the fascinating book "Zika: From the Brazilian Backlands to Global Threat" with much appreciation to its tireless author Debora Diniz. Please, in the absence of specific annotations -- infer that the interpersonal information came from her book*]

Dr. Vanessa van der Linden acknowledges *"microcephaly is only a sign, not a diagnosis"* [74], that is: there is no 1:1 association between effect and cause. Let's think about how this differs from a different birth defect/ mental retardation causation, Down's syndrome– a case of which is always caused by trisomy 21. The converse is true as well, every instance of trisomy 21 mutation necessarily brings on Down's syndrome. This is not the case with microcephaly which is not specific to any particular cause.

Investigation algorithm for microcephaly

Investigating Microcephaly

Comprehensive history and examination

Congenital, maternal or environmental factors

| Relevant investigations, e.g. TORCH | YES　　NO | Prominent other clinical features e.g. dwarfism |
| Strong suggestion of a specific disease e.g. Seckel syndrome | YES　　NO | Progressive neurology |

NO

IF YES

| Specific tests e.g. array/gene panel | Primary microcephaly | NO　YES | Neurometabolic investigations MRI Brain Targeted gene panel |

| Microarray Targeted microcephaly gene panel and/or Neurometabolic investigations | | Secondary microcephaly |

Microarray Neurometabolic investigations, MRI

| If no diagnosis is reached, perform MRI brain | | If no diagnosis is reached, genomic sequencing |

By no means a simple process to determine causality within microcephaly cases. Wasn't routinely done, Brazil, 2015, accounting for the panic's overdiagnosis numbers   Adapted from "How to support the child with microcephaly" 2018

In fact it's a very complicated and usually incomplete journey from finding microcephaly by measurement to determining a diagnosis. Most cases remain undefined for cause.[75]

When this twin tested negative for cytomegalovirus, Dr. Vanessa van der Linden (perhaps inspired by **"CHIKV, The Mission"** WhatsApp conversation) requested chikungunya testing, but the family didn't have the finances to follow through on a non-covered lab. Chikungunya had never been associated with microcephaly.

At some point in the next few weeks, on the lookout perhaps for microcephaly, she saw three cases in her hospital, and assumed they might be infection-related although testing for toxoplasmosis, "other" (parvovirus), rubella, cytomegalovirus, varicella, syphilis [the so-called TORCH(s) panel[76]] for microcephaly-associated infections had been negative. [77] At that point she consulted with her mother who got busy calling 10 regional public hospitals' neuropediatricians, et al..[78]

They had a loose idea of how many microcephaly cases occurred yearly and thought that the census at that point indicated an increase in microcephaly, which they decided to attribute to infection. Even though it's impossible to attribute any individual case of microcephaly to any one specific causation, they decided to exclude the entire pre-existing set of "TORCH(s)" panel infections already associated with microcephaly.

Instead, Drs. van der Linden took to WhatsApp to inform local MDs of a potential mystery virus' causing microcephaly, and that any and all cases be brought to them. **And, thus, it came to be:** essentially a WhatsApp echo-chamber reverberation cycle of more cases' causing more worry, alarm and panic, in turn bringing more cases. Lather, rinse, repeat.

Convinced of a novel viral cause, they queried moms, some of whom had complained of fever or rash during pregnancy; many who had not. Over the course of nine months, "who hasn't had a rash or fever?", could be a reasonable, converse question. The van der Lindens' process should not be mistaken for rigorous scientific inquiry.

In October 2015 a colleague, Dr. Adelia Souza, suggested Dr. Vanessa van der Linden discuss her thoughts with Recife's own local Zika-theorist, Dr. Carlos Brito was convinced it was Zika, despite the absence of any contemporaneous laboratory testing).

As with other parts of the 2015 Zika-microcephaly timeline's narrative, there was some institutional pushback: epidemiologists felt the Drs. van der Linden's perceived elevation of microcephaly census was not genuine, but

rather reflected previous, chronic underreporting. Dr. Brito countered by insisting that Recife's situation *"was extreme microcephaly"* -- and, irrespective of (and sweeping away) having to verify numbers, *"this was different"* – notwithstanding that later review and retrospectives would come to disprove his contentions.

Dr. Brito soon decided to check cases at the hospital, meeting with neonatologist Dr. Jucielle Menezes who showed him 16 microcephalic babies who all (not coincidentally) were under the service of Dr. Ana van der Linden. It was at this point that he made a retrospective and retroactive questionnaire, collecting stories about rash etc. from 26 local moms with microcephaly-births from the previous few months.

Dr. Brito called his friend, colleague, and fellow **CHIKV**/Zika-hunter, Dr. Kleber Luz-- who was, at the time, more skeptical. Kleber Luz eventually circled back to a letter some months prior from an Italian woman familiar with his Zika discovery who had delivered a child with birth defects subsequent to her own fever and rash. Somehow that single and signal event swayed him more than the Recife cases. Frankly, neither situation evoked actual statistical comparison and neither one should have swayed his otherwise appropriately skeptical mind.

In the meanwhile, Dr. Vanessa van der Linden's communications with her neuropediatrician brother and mother helped solidify and direct her suspicion of an external, infective cause.[79]

> *"You need to think of agents that cause epidemics, that cause many cases at the same time,"* she explained. [80]

Microcephaly already was associated with the *"TORCHS"* set of infective agents.[81] Why it ***had*** to be a new agent, in this case Zika, is unclear (except possibly through the lens of Zika-theorist Dr. Carlos Brito) – although the novelty certainly became a clarion call, noted and noticed by millions.

As chairman of Recife's state Regional Board of Medicine (CREMEPE), Dr. Brito introduced his hypothesis that Zika causes microcephaly. By coincidence, two days prior, the press had reported his hypothesis. It's unlikely anybody but Dr. Brito or an ally had informed them. He had been the only one on the scene with this theory: the direct connection between the two key initiators: Drs. Kleber Luz (Zika-origin theorist) and Vanessa van der Linden (microcephaly-increase observer).

Recife neuropediatrician Dr. Adélia Souza tapped on the brakes and accelerator simultaneously: downplaying Zika; but hedging that by confusingly evoking recent dengue -- although dengue, year after year, had never brought microcephaly. Here is her mixed message:

"It's important to stress that we cannot draw any relation with dengue, chikungunya, or Zika yet. What's happening is that since the beginning of the year, we've been experiencing a dengue epidemic, which has coincided with the gestational period of women who recently had babies with microcephaly."

There were some tensions at the medical board meeting because of both the leak to the press and its acceleration and hyping through Dr. Brito's WhatsApp group **"CHIKV, The Mission"** -- which separately had announced that **MoH** and Pan American Health Organization (**PAHO**) representatives would attend. Ultimately, they didn't, but the board was miffed anyway at the co-opting of their protocol and prerogatives: feeling Dr. Brito had put himself in the position of the tail wagging the dog: overconfident, peremptory, domineering.

Dr Brito backtracked upon being rebuked: explaining official invitations had not actually been sent. Somewhat disingenuously (presuming he had perpetrated the leak to the press) he stated, *"It was just a closed scientific meeting"*. But the state agencies had taken umbrage at being used as pawns, denying Dr. Brito a further role in policymaking on Zika or microcephaly. Dr Brito didn't skip a beat, becoming the coordinator of his own medical board's task force to monitor the growth of microcephaly in Brazil.

It's important to note that the doctors who made this connection were clinicians and not full-time academic research-scientists with the epidemiologic, laboratory and data- resources available to Brazil's major research institutions.[82] [83] [84] There was some applied skepticism from university and governmental institutional requests that mere suspicions not be released prematurely. *CHIKV, The Mission*'s and *S&SC*'s contrary motivation and approach, that it was information too important for the entire public sphere immediately not to know, prevailed.

The Pernambuco Department of Health (in Recife) broadcast October 27, 2015 the (unconfirmed) increase in microcephaly (with appropriate cautions)
– but this brought the opposite of a calming effect.[85]
Recife was already in the process of being overwhelmed with "cases" of microcephaly. More than 90% of these new visitations (in retrospect) did not reach the threshold of any microcephaly diagnosis. Vastly, these cases came from the "worried well": mothers' and clinicians' valid overconcern upon the announcement of a new mysterious pandemic, irrevocably damaging young lives at their outset.

Despite the loosened diagnostic criteria rampant within the panic, these multiples presenting to very focused and sympathetically distraught neuropediatricians clinicians fanned flames of their overconfidence in what otherwise might have seemed a very tenuous connection. Four

months later, the WHO declared a World Zika Emergency, despite only 8% of those cases' having been confirmed as likely microcephaly.[86]

Bridging the New Illness to a New Problem

Dengue -- endemic to Brazil (for either decades or centuries)[87] – had never caused a congenital microcephaly surge in Brazil or anywhere else. And that's not for lack of trying! Dengue inflicted millions of cases of illness and hundreds of deaths in Brazil alone within essentially every year prior to Zika's formulation and popularization as a possible illness in 2015.

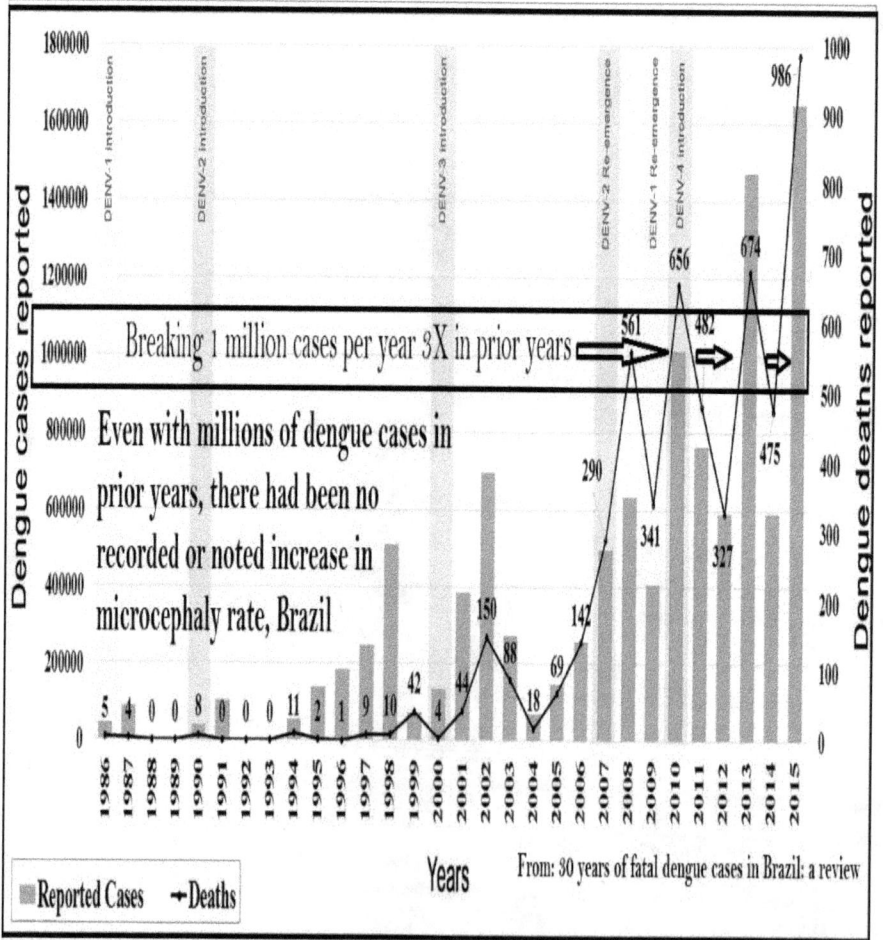

From: 30 years of fatal dengue cases in Brazil: a review

Yet, for all of dengue's huge presence in tens of millions of cases and complete absence of MICROCEPHALY, its nearly indistinguishable twin Zika somehow could, would, and did?

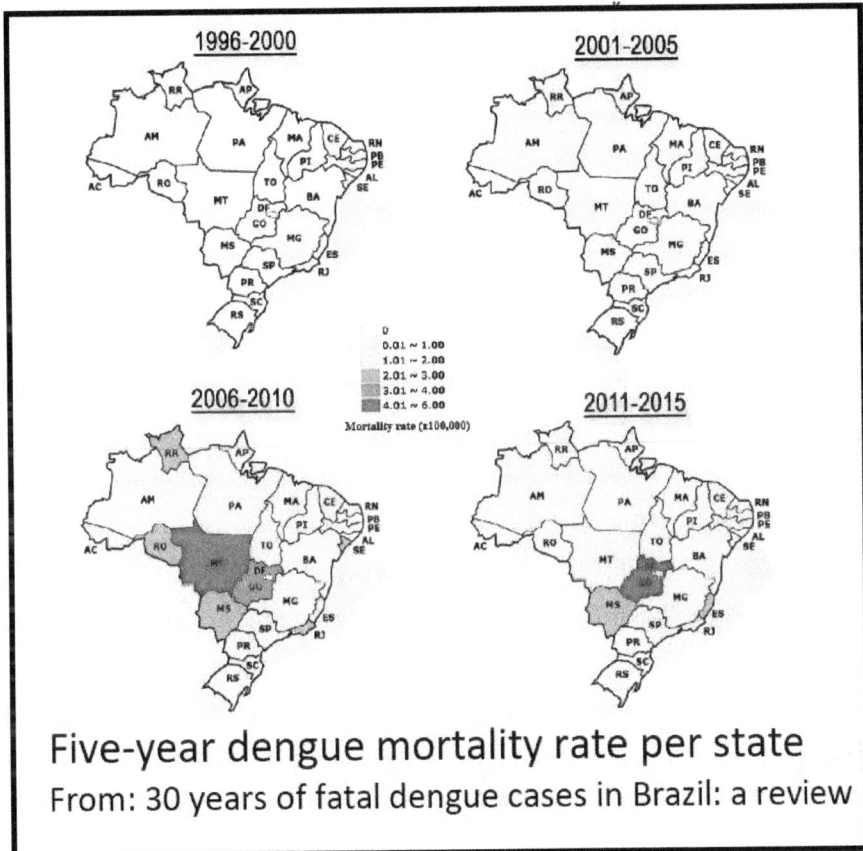

1996-2000   2001-2005

0
0.01 ~ 1.00
1.01 ~ 2.00
2.01 ~ 3.00
3.01 ~ 4.00
4.01 ~ 6.00
Mortality rate (x100,000)

2006-2010   2011-2015

Five-year dengue mortality rate per state
From: 30 years of fatal dengue cases in Brazil: a review

When, later in 2015, the Recife-area van der Linden mother and daughter pair of neuropediatricians (see inset) surmised more microcephaly-births than usual,[88] there had been no contemporaneous clinical testing for Zika in infants or their mothers ANYWHERE in Brazil.

Vanessa and Ana Van der Linden

Mother and daughter neuropediatricians from Recife, Pernambuco, who investigated a surge in cases of babies with microcephaly in the latter half of 2015.

Dr. Hélio Van der Linden (Jr.)

a neuropediatrician, He and his sister began swapping information and CT scans of babies' brains via WhatsApp and noticed an early case of microcephaly in Goainia.

It was Zika's complete novelty, rather than its kinship with dengue that led Dr. Vanessa van der Linden (coordinating with Dr. Carlos Brito) to tie

> **a Zika virus (with ZERO prior confirmed human Brazilian cases)**
>
> *to*
>
> **microcephaly cases (assigned without firm criteria -- and never shown to be at increased levels, over baseline).**

Dr. Vanessa van der Linden inferred that what she (thought she) was seeing in microcephaly-terms had to be "new" (and infectious), thus needed a "new" infectious cause, and what was the "new" infection on the block? Zika!

Dr. Brito, given the microcephalic babies' moms' contact details, queried (only) them -- for memories of rash or fever during pregnancy. He was welcome to do so, and perhaps the moms appreciated the attention, but make no mistake:

his techniques were not consistent with the scientific method. The most obvious violations are

- **"selection bias"** [89] (querying only microcephalics' and not normal babies' mothers)
- **"lack of blinding"**[90] (eliminating the buffer-layer between researcher and subject, influencing answers given to please the authoritative questioner);
- **"observer bias"** [91] (the researcher's shading answers towards his own predilection).
- The CDC's report noted that this discovery process "is subject to **recall bias**[92] and might have resulted in misclassification of potential Zika virus exposure."[93]

No firm conclusions should have been drawn from such an undertaking; certainly nothing that should have provoked talk of Zika's causing microcephaly; nothing that should have been (very soon thereafter) leaked to the press, in advance of confirmation by medical or academic research's using stricter statistical and clinical criteria. At each step, the results were deemed too important to wait. The first step, the "uncovering" of Zika as a malady (earlier in the spring) didn't stir waves as large as the announcement (that fall) that Zika was going immutably to change the lives, the new lives of babies, sadly and tragically. The steps were likely more willful than malicious; nonetheless they should not have occurred-- especially so quickly, haphazardly, and without review.

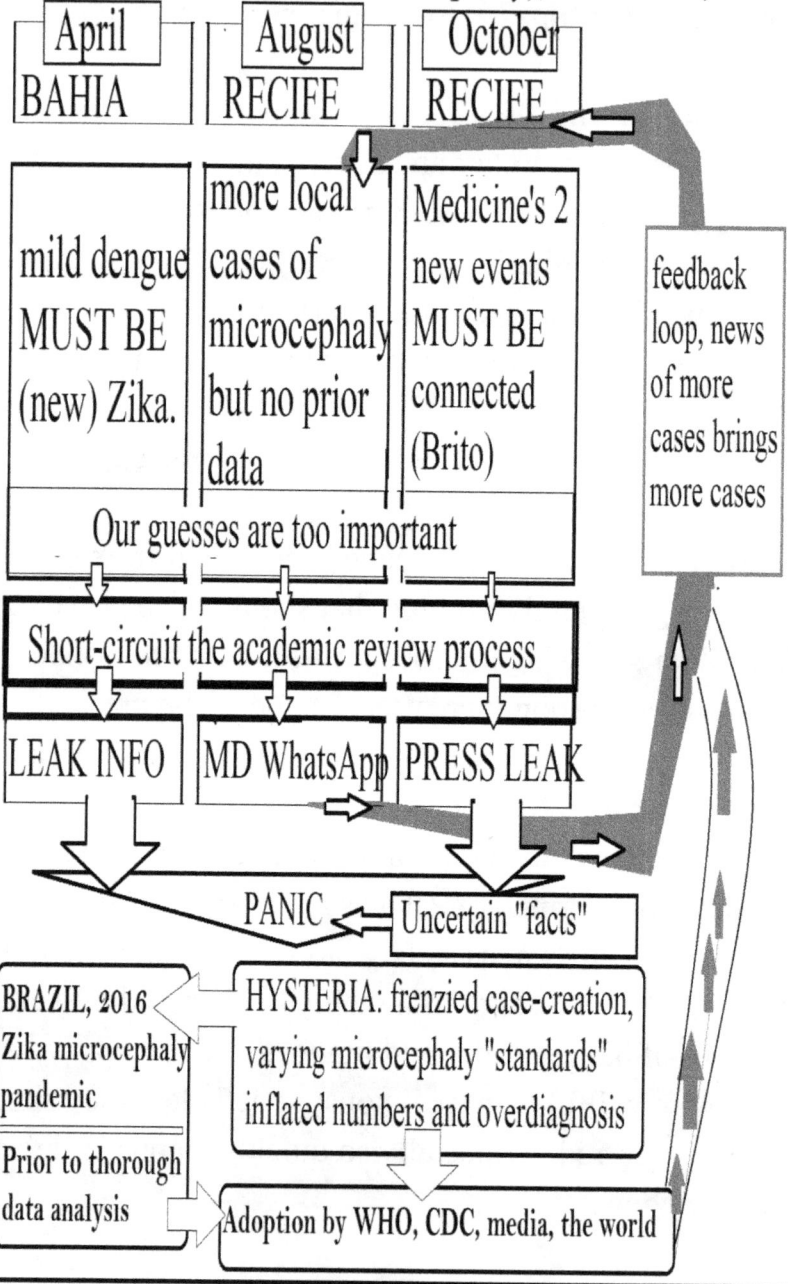

The Bridge, Tying A Potential New Illness (Zika) To A Perceived New Problem (Microcephaly), NE Brazil, 2015

April BAHIA	August RECIFE	October RECIFE
mild dengue MUST BE (new) Zika.	more local cases of microcephaly but no prior data	Medicine's 2 new events MUST BE connected (Brito)

feedback loop, news of more cases brings more cases

Our guesses are too important

Short-circuit the academic review process

LEAK INFO | MD WhatsApp | PRESS LEAK

PANIC ← Uncertain "facts"

BRAZIL, 2016 Zika microcephaly pandemic

Prior to thorough data analysis

HYSTERIA: frenzied case-creation, varying microcephaly "standards" inflated numbers and overdiagnosis

Adoption by WHO, CDC, media, the world

Intermission

If you've come this far in the story, perhaps it's time for a short break, beverage, or walk around the block. Sometimes it's all a bit overwhelming. At least that's how I have found it. There are so many threads. History (as it's occurring) is just "real life": multiple people's multiple conversations and interactions' bringing simultaneous, overlapping (and potentially contradictory) results. After the fact, it's harder to say with certainty which events caused which others. Where do we stand so far in the Zika-story?

Before delving into this matter, I had assumed (like many other medical readers with no first-hand knowledge) that the association between Zika and microcephaly had come from large-scale studies: data comparisons; verified microcephaly cases; uniform sizing-standards; retrospective analyses; long experience with the virus; documented infections followed through pregnancy; and the results' differing from control groups. I had assumed wrongly.

In fact, the proposed theory had evolved from discussions between social- or family-connected physicians at different outposts, sharing anecdotes that microcephaly had become more common, searching for a novel reason. Of course, this inchoate search for a cause met up with a similar relatively unformed recharacterization of dengue cases as completely novel Zika. And the rest as they say is "history".

What type of "history" ultimately it will be is another question. So far, it's merely the history of a strange new virus, bringing fear and panic through an incredibly sad congenital birth defect– bravely encountered and exposed by intrepid and enterprising physicians, working fairly independently, gaining insight and inspiration through each other – without much in the way of help from Brazil's research institutions or academia. So many of these qualities are admirable, and with microcephaly so heart-wrenching – a harshly different and critical interpretation of their efforts might seem impolite and unsympathetic; however, the chips must fall where they will.

And, during this "Intermission", so we don't lose sight of those potentially affected, remember those potentially affected (as pictured). Microcephaly existed before the Zika-microcephaly scare and will exist for quite some time after. Many families benefit from a special state-stipend for microcephaly, developed after the Zika-formulation. [94] There's nothing wrong with that, but this was not the most expedient pathway to helping women with birth defects, isolating and popularizing only one (microcephaly) whose connection with any new illness was, as we have seen, quite tenuous.[95]

In the absence of a Zika-microcephaly recurrence even in Brazil the following year and in any year thereafter to date -- and a similar absence of any similar occurrence worldwide in the interim, despite Zika's pervasively being tested -- alternate theories need to arise. These aren't percolating from within the entrenched and funded university-based research apparatus or academics, so it's left for similarly but contrarily intrepid and enterprising independent physicians to puzzle this out. It's certainly not impossible that -- as well-meaning and thoughtful as the

Brazil doctors on the forefront were -- that they were wrong; that their claims accepted along the way as "science" merely were the wildest of conjectures -- unproven at the time, and certainly not reinforced by later collection and/or reconstruction of the data.

To be fair to Zika-research academics, in the subsequent years of no microcephaly pandemic, there have been proposed "reasons" for the absence: theories of "mutant strains", dengue-based immunity, and other patches layered over the unworkable existing theory -- but nobody is aiming to overturn the entire Zika-microcephaly hypothesis and/or either half individually. That is, perhaps it actually wasn't Zika in Brazil (but dengue); that there may not have been any significant jump in microcephaly numbers in 2015 over baseline; that, even if either one or both were "real" -- one did not cause the other.

In the years after the panic, Brazil started asking similar questions including this one in mid-2019,*"Is Brazil still experiencing an epidemic of microcephaly?"*.

> *"Brazil experienced a Zika epidemic during late 2015 and early 2016.... Here in Brazil, the end of the national emergency for the disease was announced by the Ministry of Health in May 2017, given the reduction in the number of Zika cases. In 2017 and 2018, approximately 17,000 and 8,000 cases were registered in the country, respectively. "*

Fundação Oswaldo Cruz (**FIOCRUZ**) in Rio de Janeiro, Brazil's main public health research institution asks the question but doesn't fully answer.. Let's look more deeply and see if we can.

The Aedes aegypti mosquito tags a million or so Brazilians with dengue per year. Brazil was on hyper-alert for Zika following the pandemic yet found only 1% as many Zika-cases as dengue-. That is fairly close to not having found any at all, especially given the possibility that the Zika positives were just dengue cases' falsely showing positive given their near identical biology.

Trying to tease out Zika's true microcephaly rate and defining what should actually be considered as microcephaly is covered more fully elsewhere within *Overturning Zika* – but suffice to say there is a special subset of microcephaly which is designated "severe". Those studying the subject believed Zika's ability to generate severe cases occurred at around 1/1000 (00.1 %) of infected pregnancies, three times higher than the backdrop rate of 3/10,000 (or 00.03%) using the WHO's Intergrowth standards.

So, for **FIOCRUZ'** 2017 and 2018 Zika case numbers, we would have to know how many were in the first trimester of pregnancy , susceptible to the potential microcephaly danger. Around 2% of the Brazilian population at any given moment is pregnant, that would mean about 500 Zika-pregnancy cases TOTAL combining 2017 and 2018 – with only one third of those occurring within the first trimester. This would total out to approximately 0 to 1 Zika-attributable severe microcephaly births per year 2017, 2018 for Brazil against a backdrop of around 900 from other varied and nonspecific causes including genetics. [96]

Additionally (for these two years combined) there would be, nationwide, only around seven cases of moderate microcephaly per WHO standards and researchers' suggested congenital damage rate of 4% of first-trimester Zika -exposure births. Zika-attributable cases countable on two hands should not qualify as an "epidemic"

So the short answer to **FIOCRUZ's** mid-2019 question: *"**Is Brazil experiencing an epidemic of microcephaly?**"*, as far as Zika is concerned, is *"**NO**"*. Ditto 2020, and 2021. Frankly every year since 2016 has been "no" and we will go on to show that for 2015, well, the answer also was "no".

Checks and Balances?

We ended the last chapter with **FIOCRUZ-Brazil's** questioning whether the Zika-caused microcephaly epidemic persisted in subsequent years. Their "answer" supplied no direct answer but their nearly nonexistent later Zika tallies implied a resounding *"NO"*. ***Overturning Zika***'s addition was to question the premise: perhaps there had been no Brazilian microcephaly epidemic in the first place. Leaving that aside – stipulating and accepting the standard narrative of Zika's presence, *AND* a 2015 microcephaly spike– scientific standards should have reinforced the extreme unlikelihood of

- The mere relabeling of mild dengue as novel Zika's solving the very next local medical anomaly; and
- Adding any <u>new causation to a rare, nonspecific illness (microcephaly, already with its dozens of loose associations)</u> so suddenly --- absent thorough data comparisons or an adequate time-window of observation.

This whole topic brings up how we as a society -- or more specifically, certain bodies within fields of expertise, like "science" -- decide what is essentially "fact". Using sports as an analogy again, players protest line-calls in tennis and basketball all the time; however, the "fact" of the matter is established fairly quickly, through referees and video-review. Moving from basketball and tennis courts to judicial courts, "facts" are determined after the judge's deliberation, which itself is subject to appeals and review. It may take a little bit longer, but it's a similar overall process.

In science, there's a geographically-wider distribution of deliberation and debate -- but no "final decision"; no single referee or "judge" to determine "fact"; no final, absolute "court". Instead there's a "hive-mind" of scientists' communicating principally through journals, meetings, talks and academia -- eventually coming to an agreement (ideally after having attempted and succeeded in reproducing the results presented, through exactly the same methods, materials and circumstances).

Naturally, this confirmation process requires the new concept's lead scientists' allowing others adequate time for testing. Premature public pronouncements are not helpful. The lay press and the general public don't fully grasp this process, nor should they. That is precisely why it is so incredibly important that scientists (or those intending to act as scientists) adhere to these principles and follow the proper procedures of submitting their theories to others before "leaking" to the press – so as not to take advantage of a "loophole" of swaying popular opinion based on unsubstantiated thoughts or results, however well-meaning the discoverer might be.

Or as Shakespeare's King Lear said -- long before 1989's "science by press-release", the infamous "cold fusion" debacle -- *"O, that way madness lies; let me shun that; / No more of that"*.

Cold Fusion: the cautionary tale of "science" by press release

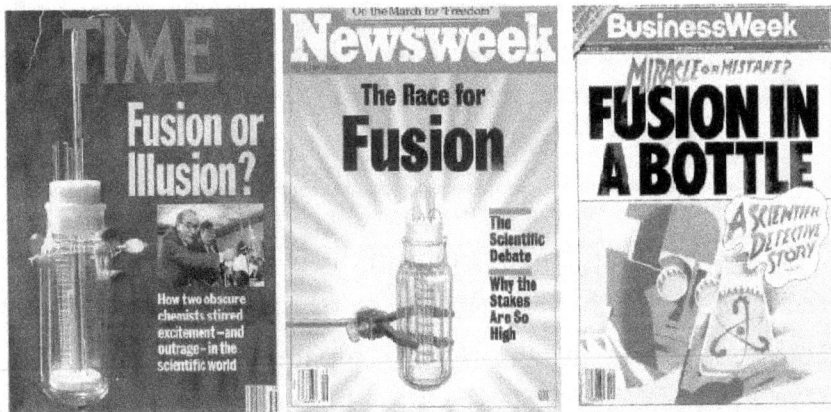

"In 1989, the scientific world was turned upside down when two researchers announced they had tamed the power of nuclear fusion in a simple electrolysis cell. The excitement quickly died when the scientific community came to a consensus that the findings weren't real—"cold fusion" became a synonym for junk science.

In the quarter-century since, a surprising number of researchers continue to report unexplainable excess heat effects in similar experiments, and several companies have announced plans to commercialize technologies, hoping to revolutionize the energy industry. Yet, no one has delivered on their promise, leaving several possible conclusions:

1. The claims are correct, but need more time to develop;
2. Those making the claims are committing an elaborate ruse; or
3. It really is junk science that won't go away."

Stephen K. Ritter
November 7, 2016 | Chemical and Engineering News Volume 94, Issue 44
https://cen.acs.org/articles/94/i44/Cold-fusion-died-25-years.html

There are certain similarities with phases of the Zika-microcephaly discovery story, competitive scientists' posturing for recognition and research funding.[97] Quickly upon announcement, before disillusionment and doubts set in, the physicist claimants' institution received $5 million in research funding earmarked for them.[98] The whole affair is now looked upon skeptically, with suspicions of fraud. Although, by this interpretation, cold fusion may not have been fraud, per se. At the time somebody wrote:

> It is important to keep in mind, Anderson pointed out, that "honest error and misinterpretation" are excluded from the definition of "fraud."

[99]

There are two instances within the Zika-microcephaly story of submitting theories directly to the press in advance of others' having an opportunity to gauge accuracy: Dr. Gubio Soares Campos decided he could *"benefit the public more"* by putting his "Zika-as-illness" thoughts in the popular press rather than having them first reviewed by other scientists. Notably when **S&SC**'s results were checked, there were numerous concerning issues found by SESAB. Nonetheless, by that time the "horse was out of the barn": people already were excited, titillated and frightened by the mysterious new illness.

During the perceived microcephaly spike, leaking the theory that Zika was the cause was likely instrumental in igniting the panic. At that point, Zika in the public imagination evolved from merely "mild dengue" into some "ticking time bomb" of microcephaly. Any mom pregnant

at the time of this news could but wonder if her imminent baby's life had already been tragically and irrevocably compromised. That made the story instantly, magnetically compelling and publicly, sensationally unavoidable (right or wrong). That this theory had had no proof, no testing corroborating it was beyond the public to know or apparently the press to investigate or emphasize. That had been the (neglected) duty of the theorist, Dr. Carlos Brito, prior to leaking (presuming it was he).

So what's the proper roadmap from an idea, to a theory, and finally a "fact"? Well, first of all, as the physical system being observed becomes more complicated, it's harder and harder to establish a "fact". We see this with a "physical system" as simple as eating eggs for breakfast. Until 1968, dietary eggs had been universally promoted for better health; whereupon they were to be avoided for their assumed cardiovascular danger, via cholesterol.

1984

1999

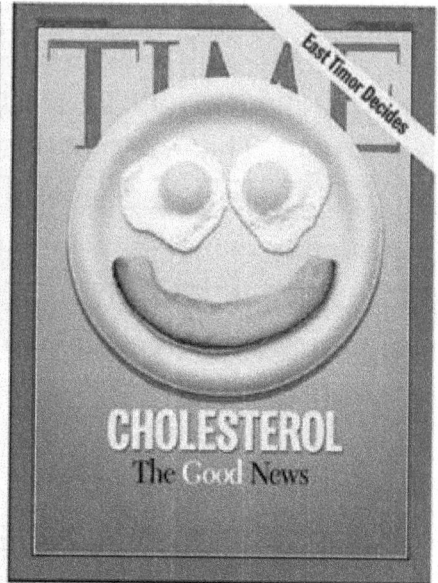

Rehabilitating the egg?

Fast-forward a few decades: eggs, now "rehabilitated",[100] are good to go: healthful and unrelated to heart issues. So what about "facts"? Which "fact" is correct? Were you correct to avoid them; is it correct to eat them daily now? And that's just a simple egg. The situation can become very much more complicated, very quickly.

Science, such as it is determined at a given moment, is often folded into public policy or becomes its driver. Anybody having experienced the evolving and varied Covid-19's dictates and restrictions needs no reminder. When eggs were in disfavor, should the government have become involved? Should they have been outlawed or been

given warning labels, like cigarettes'? Were farmers later owed an apology?

Plant biologist Prof. Peter Horton capsulizes the ideal process for making policy from an evolving science:[101]

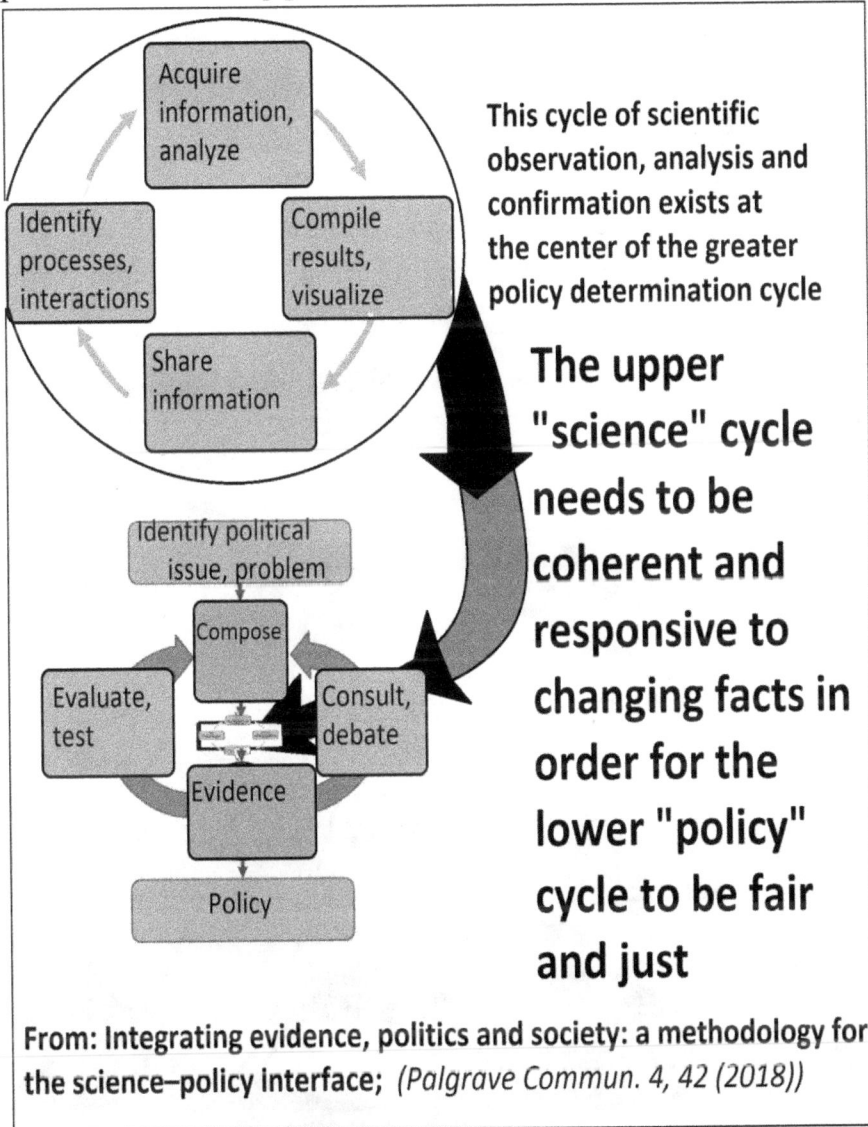

This cycle of scientific observation, analysis and confirmation exists at the center of the greater policy determination cycle

The upper "science" cycle needs to be coherent and responsive to changing facts in order for the lower "policy" cycle to be fair and just

From: Integrating evidence, politics and society: a methodology for the science–policy interface; *(Palgrave Commun. 4, 42 (2018))*

Importantly, both the inner and outer circle imply ongoing review and modifications through open sharing of all data for the inner (science) circle, and debate and deliberation for the outer (policy) circle.

Combining the dimension of time, both of these cycles approximate a spiral staircase or corkscrew, ultimately coming to a point; a destination; a steady-state understanding of the situation – including possibly that it was (as in the case of eggs) a false alarm (one hopes). Sadly, it seems that human nature in many ways stubbornly duplicates *Dr. Seuss's Zax,*[102] *in the Prairie of Prax*, not willing to budge, change directions, or accommodate – steadfastly maintaining an idea long past its expiration date.

Mathematics has its proofs, but as the scientific subject moves away from particle physics towards more complicated multi-body physical systems (increasing in complexity in each step from chemistry to biology to medicine to epidemiology to social policy), theories become more difficult to prove. This XKCD comic summarizes the "hierarchy of science":[103]

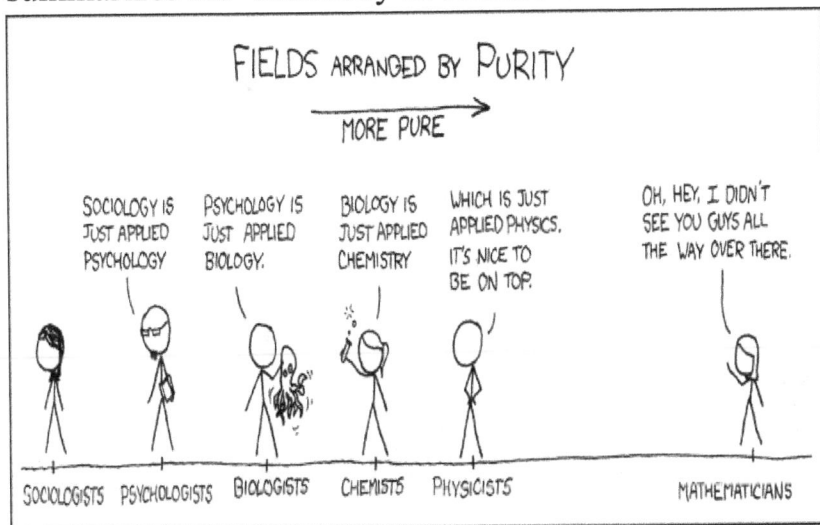

FIELDS ARRANGED BY PURITY

MORE PURE →

SOCIOLOGY IS JUST APPLIED PSYCHOLOGY

PSYCHOLOGY IS JUST APPLIED BIOLOGY.

BIOLOGY IS JUST APPLIED CHEMISTRY

WHICH IS JUST APPLIED PHYSICS. IT'S NICE TO BE ON TOP.

OH, HEY, I DIDN'T SEE YOU GUYS ALL THE WAY OVER THERE.

SOCIOLOGISTS   PSYCHOLOGISTS   BIOLOGISTS   CHEMISTS   PHYSICISTS   MATHEMATICIANS

Part of the problem is that not every medical theory can be adequately set up for experimental review, given ethical reasons, e.g. the danger to the participants from infection. In the case of Zika-microcephaly, even a skeptic would not want to bring Zika via *Aedes aegypti* mosquito to pregnant women "just to check" to see if birth defects occurred. Experimental cleverness is needed to create potential proofs or rebuttals, while circumventing human danger. Sometimes this is through testing animals, using laboratory reagents -- or comparing different geographies or demographics that may already have experienced life with (versus without) the toxin in question. For instance did Brazil's tropical neighbor Colombia have the same microcephaly statistics, post Zika? Did Brazil, itself, the subsequent year?

Science has the pretense or appearance of its facts being scientifically derived; however in reality there may be some similarity with the process, through popularity and consensus, of determining celebrity or fame – including *"we know it when we see it"*.[104] *"America's Got Talent"* and other contest shows may not seem scientific, but in certain aspects they may be more so than ***"SCIENCE"*** (the "hive-mind") itself. Every year they might pick some unlikely outlier, but overall, they separate the talented from the untalented, thoroughly, completely and reliably. The same could be said for "Dancing with The Stars" and the numerous other shows I never personally watch.

Science has its "scientific method" (pictured) with its checks and balances (which are not always completely followed in practice). Ideally,
*"In modern science, a scientific theory is a tested and expanded hypothesis (a guess) that explains many experiments. It fits ideas together in a framework. **If anyone finds a case where all or part of a scientific theory***

is false, then that theory is either changed or thrown out.

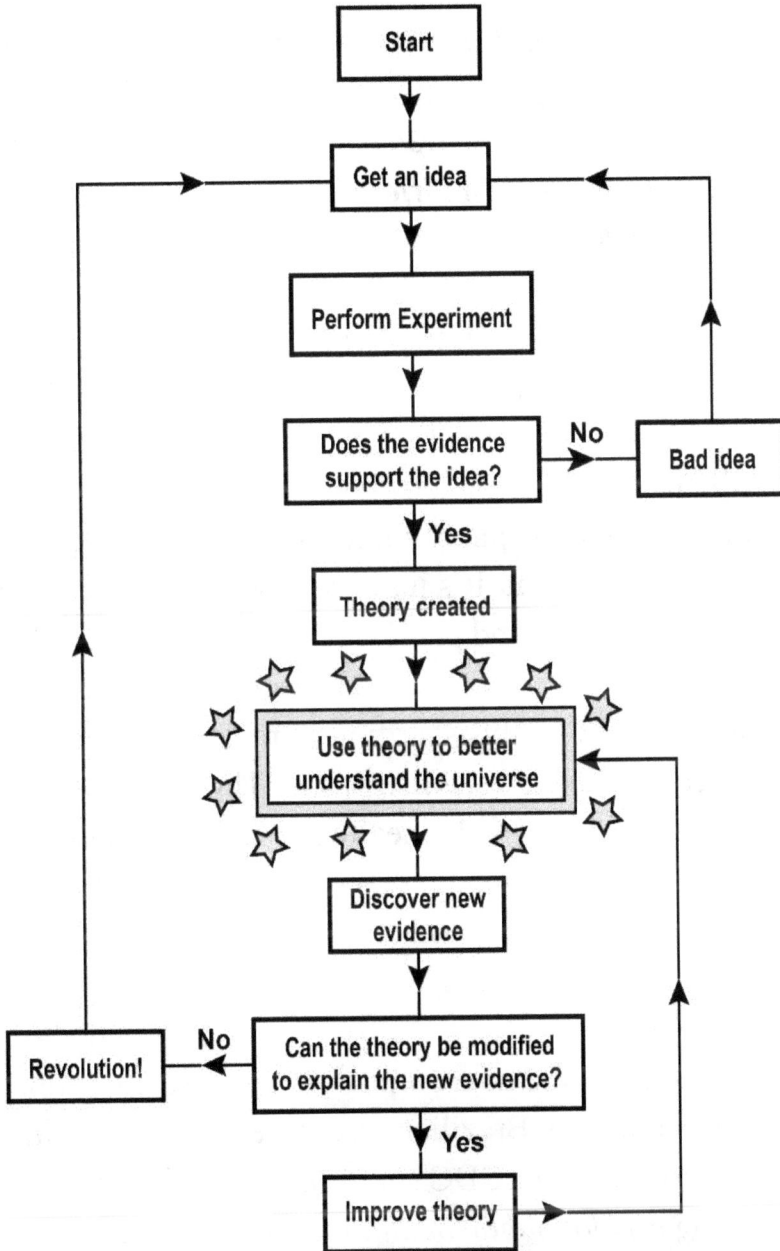

An example of a scientific theory that underwent many changes is the "germ theory" of disease. In ancient times, people believed that diseases were caused by the gods, or by curses, or by improper behavior. **To be a scientific theory, a theory must be tested a large number of times, by different scientists in different places, and must pass the test every time.** *"*[105]

There were aspects of SCIENCE's feedback loop's working as intended, with pushback on the original Zika-as-illness theory (by Brazil's SESAB). Later that same year, there were multiple challenges to the other two legs of the stool: the microcephaly spike, and the association of Zika as the cause of it. It's just that having the stories "out there" at fever pitch and panic, prematurely released, leaked to the press never allowed adequate time for the internal scientific analysis, criticism, review and theory-adjustment needed to establish "fact". This deprived any adequate footing for public health policy skeptics to "tap the brakes". The pressure to "do something" about the perceived problem was overwhelming, and eclipsed objections made at the time.

Here are some of the microcephaly-, and Zika-causation - challenges noted by Brazilian researchers, in 2015 -- in coordination with our CDC (*in its main vehicle for publishing crucial public health information and recommendations,* **the MMWR Journal**):[106]

- Prior (historical) levels of microcephaly in Brazil were approximately ten times lower than world standards (and/or underreported).
- Reporting ballooned 40-fold in 2015, suggesting a sharp increase in birth prevalence and/or increased case-reporting.
- But, actually, this was only a couple of times higher than what likely was the actual world average, given the prior under-counting.[107]
- In November 2015, Brazil's Ministry of Health (**MoH**) announced a microcephaly "special notification protocol," vastly increasing worries, concerns, and false positives
- **Before this notification, throughout Brazil, infant head-circumference had not been routinely recorded.**
- Mild cases of microcephaly might not have been reported.
- Since the **MoH** alert and the attendant media coverage of the outbreak, surveillance for microcephaly and physician-reporting of suspected cases increased.[108] [109]
- And interestingly, **once there was greater scrutiny -- the subsequent year, 2016, the (presumed) microcephaly-increase vanished.**[110]
- Zika-virus infection was not laboratory-confirmed in infants or their mothers at the time of the initial

theorized connection between Zika and microcephaly.

- Zika was ascertained *(sic)* only retroactively! There was no laboratory testing of any of the mothers of microcephalic babies during pregnancy or at birth, 2015.

- Interested physicians (e.g. Dr. Carlos Brito) questioned these very mothers after knowing of the microcephalic babies about nonspecific rash or fever and attributed these highly common symptoms to Zika (!)[111]

- Asking traumatized mothers (after the fact) about some possible mild rash during pregnancy clearly invokes recall bias, almost necessarily resulting in misclassification of events claimed to have caused exposure to the Zika virus.

Humans are susceptible to the "*post hoc ergo propter hoc*" [112] fallacy (associating one event that came after another event as having been resulting from that earlier event). It takes willpower and restraint to wait out a chain of events before making peremptory and premature conclusions.

Nonetheless, if Zika did this, we should know; *if it didn't*, we should expose the happier truth. Perhaps not immediately, but later in the due course of time — and there's no time better than the present. More than half a decade later, it should seem incredibly curious that this virus, which the World Health Organization (WHO) warned as a pandemic set to cause serious damage in

tropical regions around the globe, ultimately turned out to be such a non-event, fizzling out even within a year of the scare.

The 18th-century satirist, Jonathan Swift, stated

> *"It often happens that if a Lie be believed only for an Hour, it has done its Work, and there is no farther occasion for it. Falsehood flies, and the Truth comes limping after it; so that when Men come to be undeceived, it is too late; the Jest is over, and the Tale has had its Effect..."*[113]

It's hard to compete with popular conceptions. Science was handicapped right from the start regarding the Zika-microcephaly story given the premature release of unverified information– and that the emotional nature of seeing the extremely sad pictures of hugely compromised and limited infants, independent of whatever cause it might have been.

The Shifting Sands of Shifting Standards

Microcephaly has many potential causes and at the same time none specifically.[114] Currently all it has are loose associations, many of which (those underlined) occur more frequently in poverty-conditions (such as Northeast Brazil's).

> Disruptive injuries
> Infections: "TORCHES" (toxoplasmosis, rubella, cytomegalovirus, herpes varicella, syphilis) and HIV
> Poorly controlled maternal diabetes
> Deprivation
> Maternal hypothyroidism
> Maternal folate deficiency
> Maternal malnutrition
> Alcohol-overuse
> Teratogens: hydantoin, radiation
> Maternal phenylketonuria
> Placental insufficiency
> Death of a monozygous twin
> Ischemic or Hemorrhagic stroke [115]

The syndrome itself has also lacked absolute definition. At different times and places, the metric has varied and occasionally alternated [between -2SD and -3SD below the mean: 2.5% versus 0.15% of babies (a 17x-difference in standards!)]. Under the larger arrow " pictured, some 2.1% of babies are just smaller, without necessarily having short-circuited any possibility for intellectual growth later. Those

in the 0.15% "**-3σ**" zone (under the nearly flat left side of the curve), though, have a distinctly sadder future.

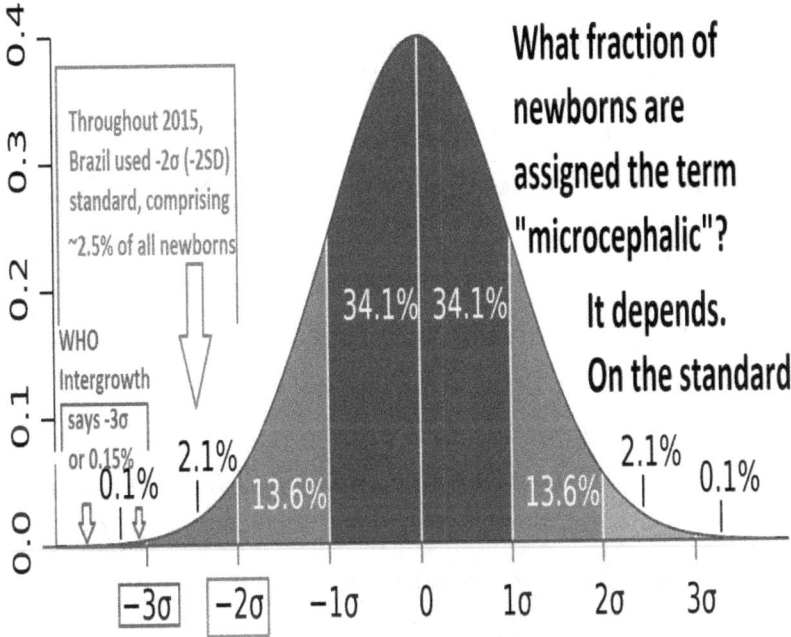

Throughout 2015, Brazil used -2σ (-2SD) standard, comprising ~2.5% of all newborns

WHO Intergrowth says -3σ or 0.15% 0.1%

What fraction of newborns are assigned the term "microcephalic"?

It depends.

On the standard

34.1% 34.1%

2.1%

13.6%

13.6%

2.1%

0.1%

0.1%

-3σ   -2σ   -1σ   0   1σ   2σ   3σ

For OB/GYNs, microcephaly diagnosis merely entails an infant's falling on one side of the locally-assigned (and potentially changing) demarcation in birth head-size; while for geneticists, "true microcephaly" is a specific autosomal recessive severe abnormality comprising about 00.01 % of births.[116] From the outset, it was implicit that Brazil's microcephaly burst was not the genetic variety.

Even when a physician obtains a good recent personal history of the pregnancy from the mother, along with laboratory testing, this is never conclusive to determine a precise cause of microcephaly. In fact the diagnostic process is fairly involved, and the protocol was not followed during panic-stricken Brazil's Zika scare.

Here's a fairly intricate and precise diagnostic protocol developed only two years prior to the 2015 scare. There is no need to memorize this diagram, but keep in mind this level of diagnostic care and clarity did not occur in Zika-panicked Recife, Brazil.

Microcephaly: nomenclature matters! **It is important to be clear that the term 'microcephaly' is a clinical finding, and should not be used as a disease designation.** *It is commonly defined as a significant reduction in the occipital-frontal head circumference (OFC) compared with age and sex (but sometimes not ethnically) matched controls.*

- **Using the label of microcephaly in the −2 SD to −3 SD population, does not help with management as a significant proportion of these children will be 'normal';**
- *Yield from genetic investigation is low;*
- *Head shape matters as a round head contains a greater volume of brain than an elliptical one of the same circumference.*"[117]

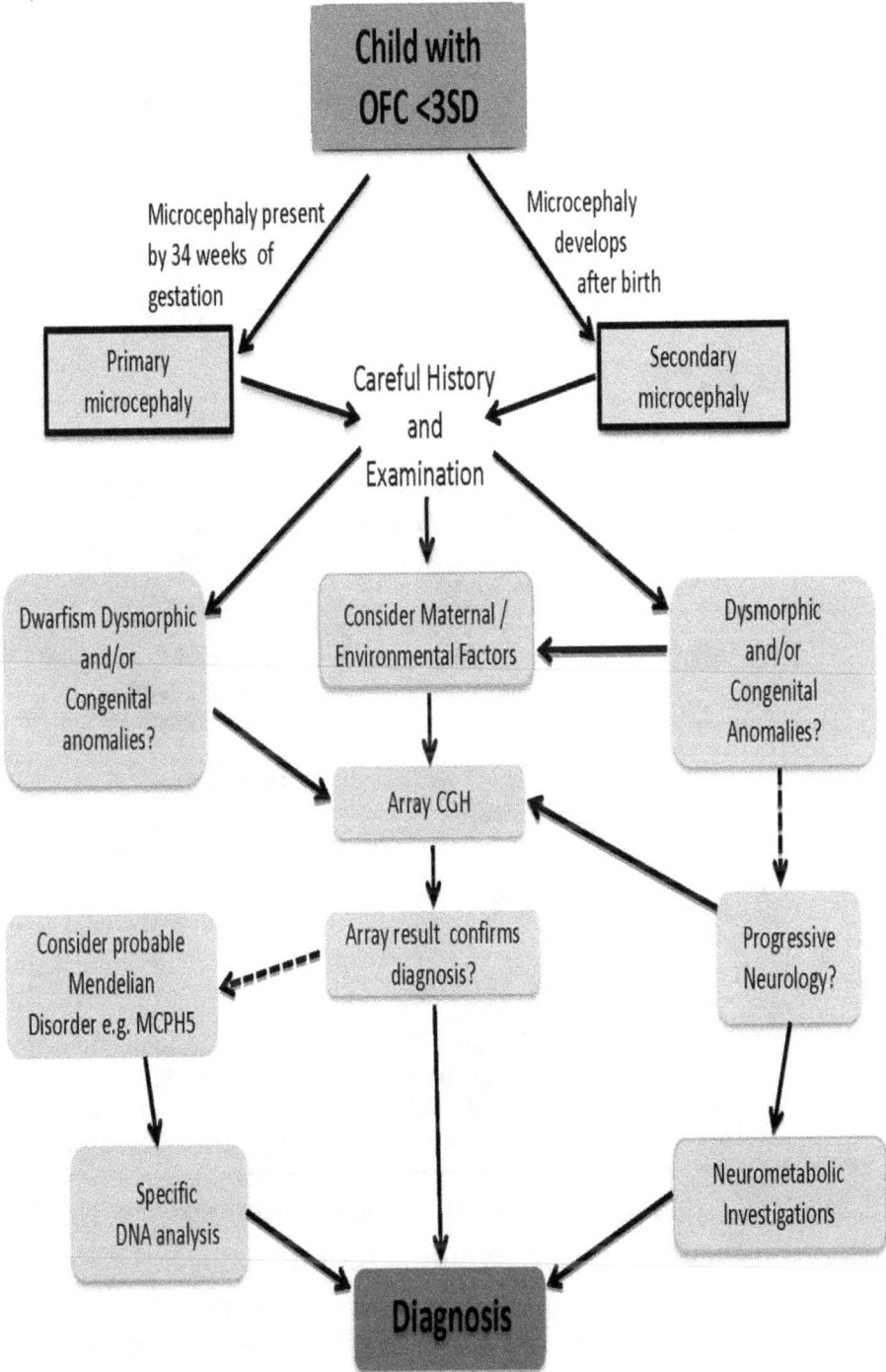

Once the microcephaly panic began, there was an overwhelming of the system evidenced by the later scaling back tenfold or more of the microcephaly claims as invalid. Further evidence is that during this same period, late 2015, Brazil changed and tightened its diagnostic standard for microcephaly determination[118].

From "Microcephaly in Brazil: how to interpret reported numbers" Case numbers varied drastically, depending on cutoff standards.

Brazil's lax microcephaly standards before 12.2015 magnified MC cases. Using the international standards to measure microcephaly dropped (estimated) case numbers 200 -fold

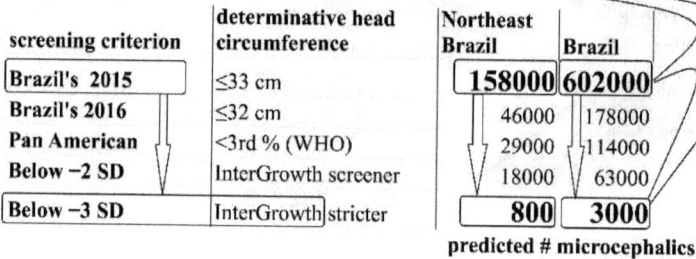

screening criterion	determinative head circumference	Northeast Brazil	Brazil
Brazil's 2015	≤33 cm	158000	602000
Brazil's 2016	≤32 cm	46000	178000
Pan American	<3rd % (WHO)	29000	114000
Below −2 SD	InterGrowth screener	18000	63000
Below −3 SD	InterGrowth stricter	800	3000

predicted # microcephalics

"This temporal increase in suspected cases of microcephaly could also be distorted given both raised awareness, with more children than usual being measured and reported, and changing definitions of microcephaly over time. These results must be interpreted with caution given the small number of cases and the possibility of notification bias. The table shows that the number of cases of suspected microcephaly range from more than 600000 with the initial criterion from the Ministry of Health to just over 3000 if -3 SD is used as the cutoff. " The Lancet, February 5, 2016

In neither scientific nor journalistic reporting of the event was this change of rules featured prominently.

For instance, in this very thorough epidemiologic review, *"The Epidemic of Zika Virus–Related Microcephaly in Brazil* (Am J Public Health. 2016 April)" -- week by week numbers are given, with no mention of the contemporaneous change in Brazil's microcephaly-standard.[119] This could, of course, mean that "on the ground" the diagnosis was continuing apace by whatever standards (or lack thereof) were operative at the time. There is a tacit acknowledgment as well of "overreporting", which likely resulted from a combination of over-concern and overdiagnosis, hand-in-hand with the ongoing panic.

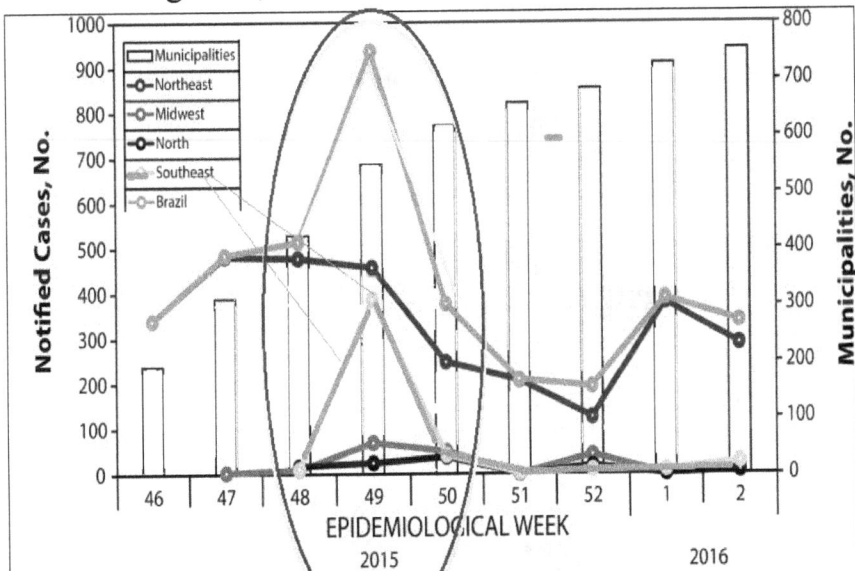

Brazil's largest microcephaly spike was a reflection of panic in the southeast, far from the Zika epicenter. Almost immediately it dropped. Northeast numbers had been steady, spiking previously.

from: The Epidemic of Zika Virus–Related Microcephaly in Brazil: Detection, Control, Etiology, and Future Scenarios Am J Public Health. 2016 April

The background bars represent total number of towns reporting, cumulatively

Witness the cases' peak in week 49, an order of magnitude higher than its cusps in southeast Brazil, far from the Recife/Northeast Brazil epicenter of both reporting and Aedes aegypti mosquitoes. If the phenomenon of infection had occurred broadly some twenty-to-thirty weeks earlier (amongst women randomly distributed temporally within their first trimester of pregnancy), then there would have been a blunted, wide distribution, not this spike Only a tiny fraction of these claims was maintained as microcephaly, proving the point of overdiagnosis during panic.

Here's another indicator, this "bubble" of Brazil's Google "microcephaly" searches tracks with its claimed cases .

Search term: "Microcefalia" within Brazil

46   **"week 49" 2015**

There was no contemporaneous research, no Zika testing, no case-confirmation, no uniform standard, and no comparison with previous years' data—and therefore no factual confirmation that the clinicians' *feelings* of higher numbers of microcephaly cases were genuinely statistically

significant. It was not until late 2015 that Brazil's Ministry of Health established a nationwide microcephaly registry.[120] It was therefore impossible that Recife's neuropediatricians could have concluded any microcephaly increase.

To see the statistical significance from events happening overall at a very low frequency requires a very high sampling volume.[121] This is one of the advantages of so-called "big data": huge computer-databases can be sifted and analyzed for small associated variations' impinging upon one another. For instance, somewhere out there somebody probably has studied the best color choice for bathroom fixtures in reselling a house;[122] the best outfit to wear to an interview; the best collection of courses and activities to get into a particular college; the single best player out of thousands for a specific team to draft. But all of these require large amounts of data, and long analysis and comparison, along with retesting for validity. All the "big data" examples given are fairly common events or transactions. By contrast, the underlying rarity of microcephaly makes the discernment of small changes to microcephaly rate much harder. If it were claimed that rainfall in the Sahara changed one year (because of, say, a volcano's erupting in Indonesia), even some observed, minor change, perhaps 3.2 inches of annual rain versus 3.9, would not prove the case. It would be hard to show that that change was meaningful, and even harder to associate it with some new cause, particularly if it had only occurred in that one year and never thereafter—in the Sahara, or anywhere else.

The Microcephaly Bubble

Certainly there was a bubble of microcephaly-claims' coinciding with (or induced by, and then fanning) panic. Under closer scrutiny most all those 2015 claims dissolved, lacking genuine microcephaly characteristics. At the time the WHO was declaring an emergency out of Brazil's ~4000 reported microcephaly cases, fewer than 10% had been confirmed – and only 00.15% (a total of six cases) had been (even tenuously, after the fact) linked to the mother's possible Zika-infection.[123]

Very few in the press covered any Zika-news skeptically in 2016, so respect is given to Vox' Julia Belluz, reprinting her thoughts here (*with my additions in italics*).[124]

1. **An awareness bias** When doctors began to sound the alarm last year over the impact Zika might have on the brains of fetuses, Brazilian health officials massively ramped up the effort to find babies who may have been harmed by the virus. People were freaked out, and surely so were doctors. In that context, doctors were looking for microcephaly and may have been more likely to report the birth complication.
2. **An incorrect baseline.** There's some suspicion Brazilian doctors underreported microcephaly before Zika's arrival. *[BUT, on the basis of a standard that*

was too lax and in 2015 changed to one stricter]. All of that makes it difficult to determine if the uptick in microcephaly is real. [*that, along with no prior-year data, registry*]

3. **Difficult diagnosis**. Microcephaly is difficult to diagnose. It's a condition that's caused by a range of things — from maternal alcoholism to Down syndrome. The World Health Organization noted that the case definition for Zika needs to be standardized, since right now estimates might be capturing a lot of noise.

4. **Confusion over suspected and confirmed cases.** We in the media sometimes fail to appreciate or communicate that initial reports of infectious disease numbers don't always represent real cases.

It was a few years more before attempts to reconstruct prior-year data came to fruition[125]. Shockingly, no true increase in microcephaly from within the 2015 "pandemic" year was observable, when reviewed retroactively. Even more shockingly, this didn't become major news, itself.

FROM: Microcephaly in north-east Brazil: a retrospective study on neonates born between 2012 and 2015. The WHO neonates with microcephaly in Paraíba, Brazil, 2012–2015 Zika's numbers show no real difference from prior years'

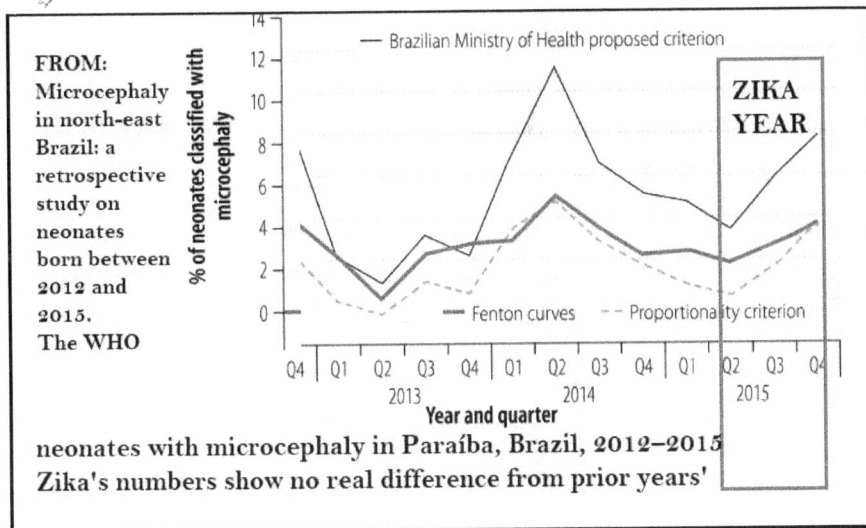

Brazil's prior years' microcephaly-rate "baseline" from 2013 and 2014 is really no different from the boxed ZIKA YEAR's . The three different lines refer to different diagnostic criteria.

Additionally, there's one glaring error (of omission) even in this report that would further lower Brazil's overall microcephaly numbers. These data were obtained from state hospitals, principally serving poverty-areas. This important, potential selection-bias fact was not disclosed in the article, but rather in a later interview with Norman Swan, quoted below.[126] Additionally, there is no way to confirm cases years after birth, with physicians and patients widely dispersed.

Moreover, microcephaly was not as well-tested as people at the time assumed. The data was poorly collected and organized, and the methods of collection were inconsistent between different locations. This report's author Dr. Sandra da Silva Mattos at the microcephaly-claims' epicenter, Recife, states that **in Brazil prior to 2016, *"the reporting of microcephaly was neither compulsory nor had clearly defined criteria"***[127] and, while there were some existing measurements regarding births, *"the original data set did not include head-circumference data."*

Therefore in actuality in 2015, when mothers throughout Brazil panicked at the prospect of Zika's causing increased rates of microcephaly, there was no scientific basis for the underlying claim. Even if we stipulate that the Drs. van der Linden in fact had noted more microcephaly cases locally in Recife than their own usual rates, that alone can't prove an epidemic. Unusual events occur, well unusually -- without upsetting the underlying order. We've all experienced winning streaks or two weeks of rain, while understanding these come unattached to any deeper meaning.

Once the Zika-microcephaly alarm was sounded there was the additional factor of "awareness bias" described above: people coming from areas surrounding Recife referred and directed to see the known expert(s). Physicians and parents understandably follow a pattern of caution at the risk of over-diagnosing, so as not to miss a potential case. They are not to blame for "racing out of the theater" when somebody else has already yelled "fire"

Those pressing for Zika/microcephaly- confirmatory data can point to more cases of ***severe microcephaly*** in the Zika-year, but these numbers represent a vanishingly small number of neonates overall, around 00.1 %.[128] Standalone, this is difficult to interpret, again with the absence of any multi-year prior data for comparison. Additionally, very small numbers have more "noise" variation, statistically.

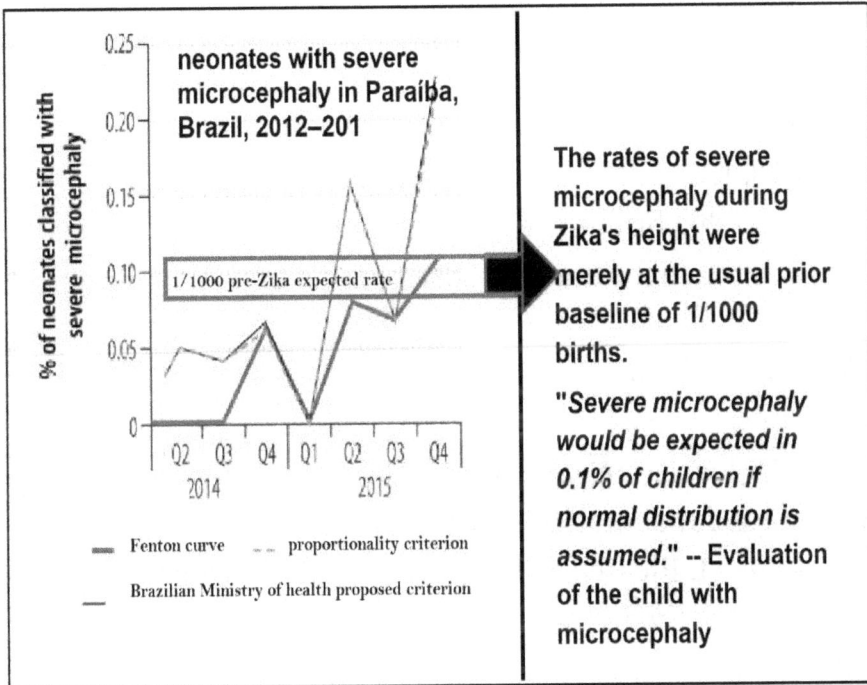

neonates with severe microcephaly in Paraíba, Brazil, 2012–201

The rates of severe microcephaly during Zika's height were merely at the usual prior baseline of 1/1000 births.

"*Severe microcephaly would be expected in 0.1% of children if normal distribution is assumed.*" -- Evaluation of the child with microcephaly

— Fenton curve   – – proportionality criterion

— Brazilian Ministry of health proposed criterion

These researchers should be applauded for even the imperfect attempt to re-create a Brazilian federal microcephaly registry, but in a sense, it has been "too little too late": the news has not been disseminated, and there has been no enforcement towards a retraction of the original claims. The physicians on either side of the

Zika/microcephaly connection hypothesis, those who found or conjured Zika (without testing) amidst a pre-existing sea of indistinguishable dengue – and those who found microcephaly changes when there were none (and attributing those misperceptions to the virus du jour) are still able to ride this wave of fame and appreciation for their presumed humanitarian, selfless acts.

And all of this happened at considerable expense of human misery. Often times when we look at the "science" and theories, we lose sight of the incredible impact and upheaval this brings to the actual population experiencing these declarations, and the fear brought on.

Dr. Sandra da Silva Mattos, the author of this retrospective view of microcephaly, interestingly came upon this topic and or information serendipitously, having done studies of newborns for completely different reasons, as a pediatric cardiologist. Please read her February 2016 interview with Dr. Norman Swan.[129]

> **Norman Swan**: *Sandra, as a pediatrician and as a citizen and as a woman actually, what's the mood in the country?*
>
> **Dr. Sandra da Silva Mattos**: *Oh, it's bordering on the panic state for pregnant women. I have seen wealthier women having moved to live in other towns further south where (there isn't) such a dramatic*

problem. I have seen women being worried whether they can get pregnant. **I've seen women using three additional pieces of clothes, one on top of the other, hoping not to be affected by that**. *And I have also been worried with women using a tremendous amount of (insect repellent) protective oils in amounts which I think may generate another problem because we don't know how much all these chemicals that they are putting on their skin may or may not also affect them…*

Norman Swan: *So in other words you are worried about pesticides on women's bodies, that that may affect the babies.*

Sandra da Silva Mattos: *Absolutely, it may be another thing because some of them are putting it on really hourly, and as much as they are taught to be cautious and not to be using it that much and all this, they say, well, between risking this causing some problems or having a baby with microcephaly, we might as well just keep putting these lotions on every two hours or every three hours.*

Norman Swan: *And we've heard of the demand for terminations (abortions) in a Catholic country.*

Sandra da Silva Mattos: *Yes, that's another enormous issue. I haven't seen anything happening directly in that area but certainly there has been talk, there has been people asking,* **and there have been people advising on at least not getting pregnant until this is better clarified.**

More than half a decade later, worried citizens deserve to have the situation *"better clarified"*. ***Overturning Zika*** aims to help in that regard.

Mosquitoes Know No Borders... The Net Result?

If 2015 was a whirlwind opening year for the Zika-microcephaly theory, early 2016 represented a tornado. Observe these BBC headlines:

B B C NEWS

Zika virus: Up to four million Zika cases predicted

Zika: Brazil to deploy army in fight against virus

Zika: Olympics plans announced by Rio authorities

Zika virus could become 'explosive pandemic'

Zika virus triggers pregnancy delay calls

Why Asia should worry about Zika too

Above, BBC headlines from the Zika-microcephaly news peak January 2016. In the subsequent months, then years – not so much: certainly no retractions of the dire predictions.

It's understandable if the omnipresently terrifying media coverage) had people convinced that all three underlying aspects

1. *Mild dengue WAS Zika*
2. *There was MORE microcephaly*
3. *One had caused the other*

were fact *(albeit with #1 and #2 unproven and with no documentation -- and #3's depending on #1 and #2).*

Nonetheless, the succeeding months of 2016 represented the theory's first "test case"-- and as such should have proved both deliberative and confirmatory (if real). There was unavoidable global public awareness under the banner of the *WHO's International Zika Public Health Emergency*[130] ("*PHEIC*") -- along with the onset of the possibility of real-time laboratory testing for Zika; as well as firmer and more standardized metrics for determining actual microcephaly; both facilitated by massive international funding.

The predictions were very rigorously, thoroughly and (seemingly) "scientifically" derived. Here's one set's predicting 2 million Zika-affected births for 2016: potentially representing 200,000 additional microcephalic babies within Latin America.[131] For reference, to the entire Zika-microcephaly panic/pandemic (2015-2016), there was attributed a total of only 1600 such cases [132])). These predictions represent a 100-fold increase over the prior

year's and basically an infinitely larger proportion than actually occurred (near-zero) through the subsequent years.

"It is difficult to make predictions, especially about the future".
– Danish Proverb

Projections of Zika virus infections & microcephaly in the first-wave epidemic.

Zika-exposed area	childbearing women	Births, annually	microcephaly cases predicted	at CDC's 5% microcephaly-rate
Brazil	*37,400,000*	*579,000*	28,950	
The Americas	93,400,000	1,650,000	82,500	

These predictions were in Nature Microbiology July 25, 2016.

If taken at face value, Brazil would be having 30,000 microcephalic births per year. Brazil's actual number was fewer than 3000 total across not one but three years of greatest scrutiny (2015, 2016, 2017). Moreover, numbers were closer to zero across the board throughout rest of Latin America.

from; "Model-based projections of Zika virus infections in childbearing women in the Americas"

These dire predictions were in keeping the premise that Zika had been able to cover large swaths of Brazil within a year or two, literally on the wings of South America's pre-existing mosquito vector, *Aedes aegypti*.

Spatial diffusion of Zika 2014 to 2016

2016

2015

2014

"The Zika Virus Epidemic in Brazil" Int J Environ Res Public Health. 2018

To those accepting this rapid Zika-transmissibility premise, the vistas for its further infesting and conquering the Americas were wide open. *Aedes aegypti* had already reached every tropical outpost on earth. All Zika "needed" was for its unwitting mobile host to continue doing its bloodthirsty "mosquito-thing": biting, drinking, and spitting its way along to passing Zika, the way it had already with dengue. Moreover the other half of Zika's symbiosis, humans, could do the rest: via sexual contact and free travel. After all, the Zika-theory requires a jump from Yap to Brazil, posited here as a Polynesian canoe-racing team's doing.[133] Any travel from the Americas to other continents would in short order produce similar

jumps, at the very least to tropical areas where *Aedes aegypti* thrives.

Here is *Aedes aegypti's* "worldwide distribution network", same as it always was, throughout the tropics, worldwide.

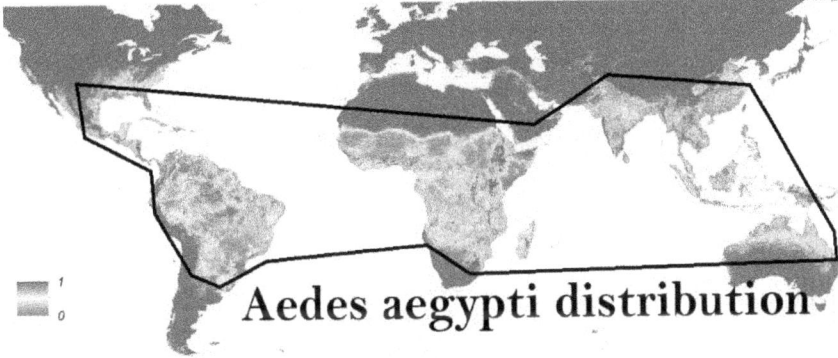

Aedes aegypti distribution

As vivid and frightening as the 2016 protections were, in one way, there were hints of a scenario even worse than this tropical "worst-case": that is, the potential for Zika to move to temperate areas as well. Much as Zika has its biologic, near-twin dengue -- *Aedes aegypti* has its close relative *Aedes albopictus*.

A tale of two mosquitoes

A. aegypti	A. albopictus
bites primarily humans (anthropophilic)	bites primarily wild and domestic animals (zoophilic) but also humans
tends to bite indoors	tends to bite outdoors
feeds multiple times per cycle of egg production	feeds once per cycle of egg production
adapts well to human urban settlements	inhabits rural and urban areas

A. albopictus is considered to have lower vector capacity than A. aegypti for transmitting arboviruses (viruses transmitted by insects), including Zika; however A. albopictus has been shown to be able to transmit Zika virus in Africa and in laboratory settings From WHO, Euro region 's white paper: "Competence of Aedes aegypti and albopictus vector species "

Aedes albopictus' terrain coincides with some of the earth's most populous zones, potentially doubling Zika's reach and microcephaly cases.

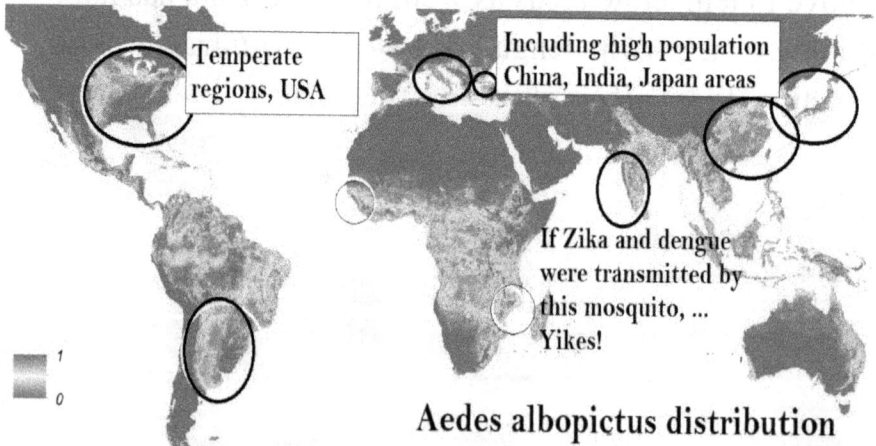

Temperate regions, USA

Including high population China, India, Japan areas

If Zika and dengue were transmitted by this mosquito, ... Yikes!

Aedes albopictus distribution

Fortunately *Aedes aegypti* and *Aedes albopictus* have their differences. While both able to carry and transmit Zika

(and dengue), tropical *Aedes aegypti* has the greater "hunger" and propensity to do so. Not taking any chances, the WHO's European branch produced *Aedes albopictus* warnings.[134] [135]

Were these temperate-zone fears' worst-case scenario outcomes realized? Fortunately, ***"no":*** there had been some worries as travelers returned from Brazil potentially with Zika in their systems– uncertain if sexual transmission of the virus could or would occur, perpetuating the spread absent any mosquito vector at all. Zika never took hold in Europe as a native illness, with or without the mosquito vectors. Working back from first principles, it should already have been obvious that ***Zika wouldn't make this crossover -- because (its near-twin) dengue hadn't.***

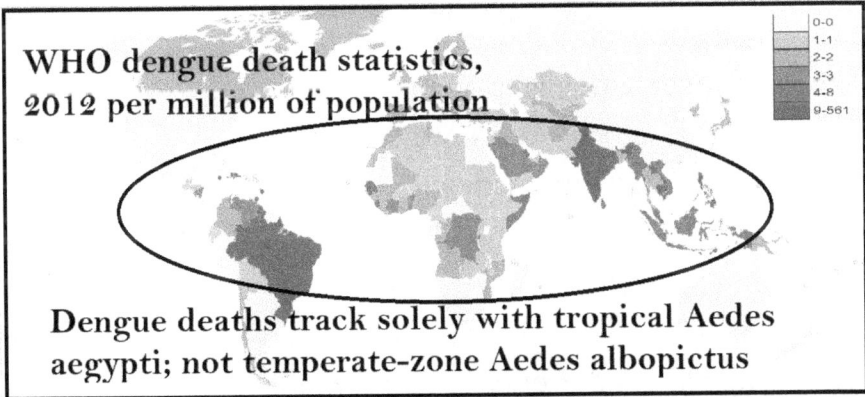

WHO dengue death statistics, 2012 per million of population

0-0	
1-1	
2-2	
3-3	
4-8	
9-561	

Dengue deaths track solely with tropical Aedes aegypti; not temperate-zone Aedes albopictus

Even better news: the microcephaly rate never changed in Europe with its stricter criteria of measurement and diagnosis, start to finish[136]. Meanwhile, the actual microcephaly rate in the continental United States appears to have remained essentially unchanged throughout this period, as well.

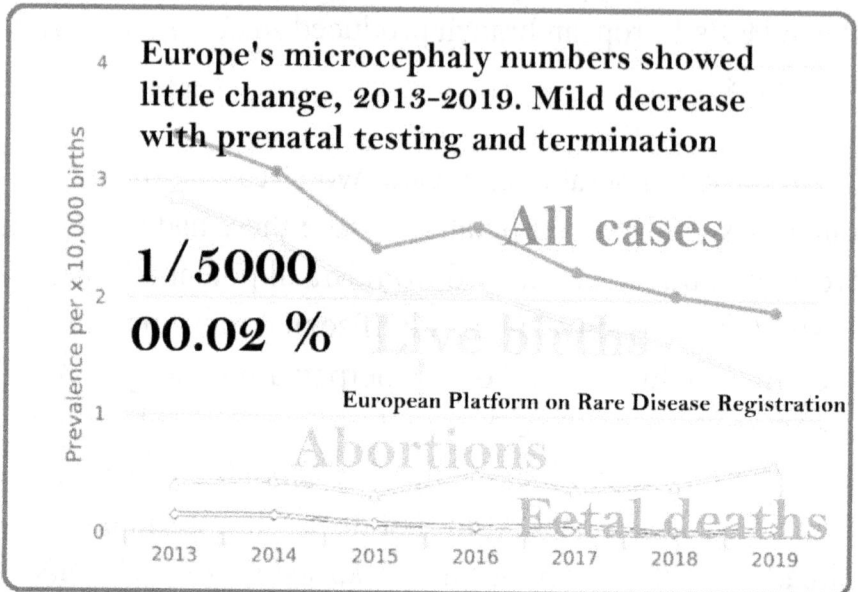

Europe's microcephaly numbers showed little change, 2013-2019. Mild decrease with prenatal testing and termination

1/5000

00.02 %

All cases

Live births

European Platform on Rare Disease Registration

Abortions

Fetal deaths

(y-axis: Prevalence per x 10,000 births; x-axis: 2013 2014 2015 2016 2017 2018 2019)

USA national data is hard to come by. Cases are tracked by individual states and later to be compiled by the CDC's U.S. Zika Pregnancy Registry (USZPR, created 2016); however, this agency has since closed shop. It was created during a declared emergency that never matched (in reality) the initial fears, so presumably as its mandate's rationale disappeared, so too did its funding.

While the US Zika Pregnancy and Infant Registry (USZPIR) existed, it showed some short-term elevation in birth defects briefly, late 2016, in the US' tropical territories (e.g. Puerto Rico), but not stateside with any statistical significance. There was a short-lived 20%

increase but only in "certain regions of Florida and Texas").[137]

Overturning Zika's multiple queries for individual southern states' microcephaly data since that time have gone unanswered. "No news is good news" likely reflects the microcephaly situation, as notably the US' overall Zika-case numbers went from negligible down to zero these last few years. US Zika numbers were never high, stateside -- and, such as they were, may just have been an artifact of media hysteria and the resulting wave of panic. Claims skyrocketed for one year coincident with the rush of Zika stories, early 2016. In the following months, with Zika-microcephaly a nonevent even at its epicenter, US claims disappeared.

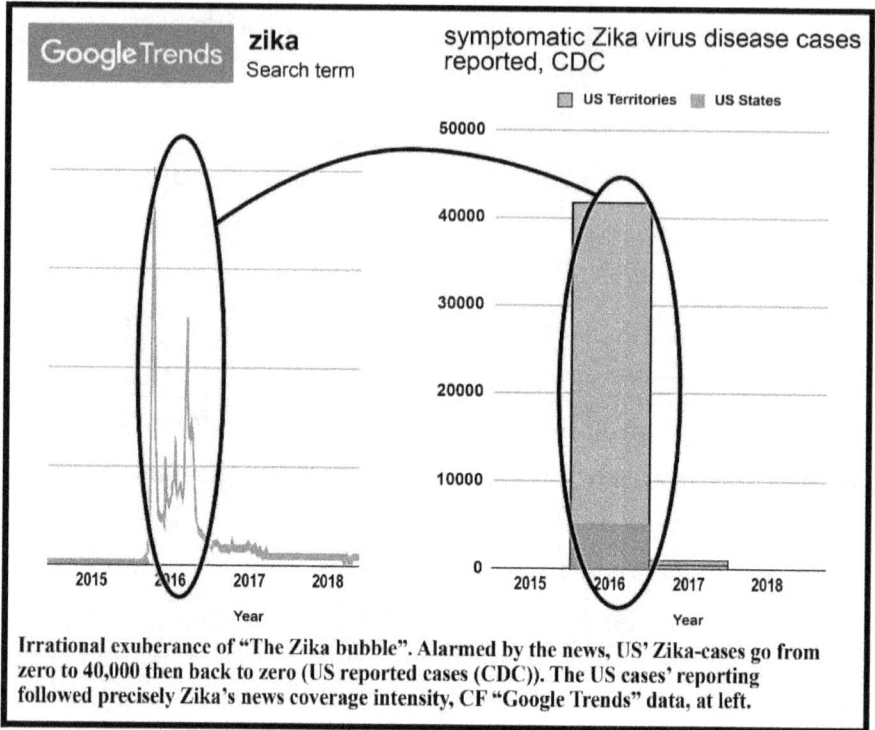

Irrational exuberance of "The Zika bubble". Alarmed by the news, US' Zika-cases go from zero to 40,000 then back to zero (US reported cases (CDC)). The US cases' reporting followed precisely Zika's news coverage intensity, CF "Google Trends" data, at left.

There is some evidence that a similar phenomenon (of claims-made aggrandized by panic) took place in Brazil. There are two different comparisons to be made:

- **Between states within Brazil,** in the origin year (2015). Those closest to Recife's breaking news had many more microcephaly claims than even those with higher rates of Aedes aegypti mosquito prevalence.

- **Between Brazil and other tropical countries within the region** – particularly those sharing borders, e.g. Colombia -- the following year, when everyone's antennae were up.

Between states within Brazil:

There were huge differences between *Aedes aegypti* - endemic Brazilian states in RATE of microcephaly-incidence.

> *"the estimated absolute risk of a predicted confirmed microcephaly case in a baby born to a woman infected during pregnancy, assuming a 50% infection rate, ranged from 0.006% in Paraná State in southern Brazil to 0.99% in Sergipe State in north-east Brazil"*

That is a 165-fold differential in microcephaly-rate. This (rather busy) chart overlays the claimed and confirmed rates of microcephaly, Brazil, 2015 labeling the regions and showing distribution of the vector mosquito. Comparisons are best made between areas with similar levels of tropical weather and *Aedes aegypti*. Notably, the North and Central West have far fewer microcephaly claims than the Northeast. *A*.

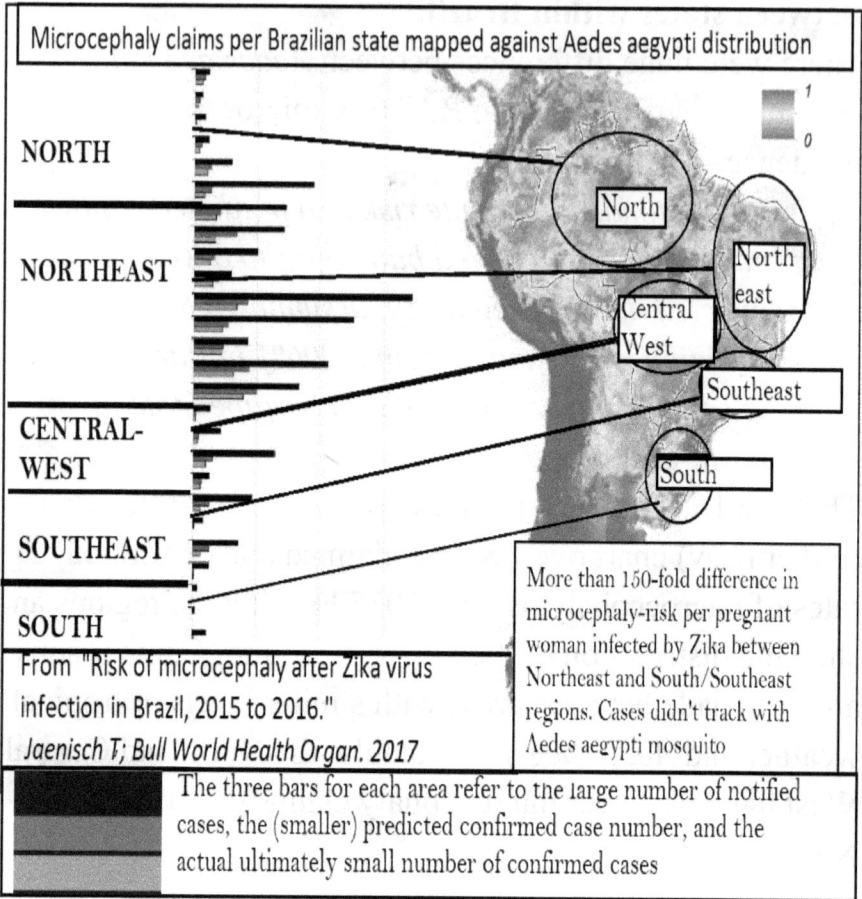

Microcephaly claims per Brazilian state mapped against Aedes aegypti distribution

NORTH

NORTHEAST

CENTRAL-WEST

SOUTHEAST

SOUTH

From "Risk of microcephaly after Zika virus infection in Brazil, 2015 to 2016." *Jaenisch T; Bull World Health Organ. 2017*

More than 150-fold difference in microcephaly-risk per pregnant woman infected by Zika between Northeast and South/Southeast regions. Cases didn't track with Aedes aegypti mosquito

The three bars for each area refer to the large number of notified cases, the (smaller) predicted confirmed case number, and the actual ultimately small number of confirmed cases

An attempted argument to explain away the huge differential might be that (somehow) Zika existed only in the Northeast; however Zika's presumed port of entry from French Polynesia was Rio de Janeiro (via the August 2014 canoe races),[138] so if it was anywhere in Brazil (2015) it would have been at least equally in Brazil's South/Southeast as its Northeast.

Between Brazil and other tropical countries within the region:

To the extent Zika was accurately measured or measurable in the subsequent year, 2016, it was tracked throughout the tropical portions of the Americas, yet (again) there was a huge differential in microcephaly rate between Brazil and every other referable nation.[139]

The Washington Post
Democracy Dies in Darkness

The Americas

Scientists are bewildered by Zika's path across Latin America By Dom Phillips and Nick Miroff
October 25, 2016

DOM. REP. 10
PUERTO RICO 2
MARTINIQUE 12
FRENCH GUIANA 10
HONDURAS 1
EL SALVADOR 4
COLOMBIA 46

Places with confirmed birth defects associated with Zika

Source: Pan American Health Organization, data as of Oct. 20.
THE WASHINGTON POST

BRAZIL 2,033

Brazil (specifically its northeast) was the only place in the hemisphere with significant increase over baseline rates of "congenital syndrome cases associated with Zika" (i.e. microcephaly).

Zika cases in the Western Hemisphere

Zika cases — Confirmed congenital syndrome cases associated with Zika

Brazil	310,061 (2,033)
Colombia	104,619 (46)
Venezuela	60,791 (0)
Martinique	34,457 (12)
Honduras	31,933 (1)
Guadalupe	30,969 (0)
Puerto Rico	29,084 (2)
El Salvador	11,315 (4)
French Guiana	10,273 (10)
Jamaica	6,377 (0)
Dom. Rep.	5,203 (10)
Mexico	4,837 (0)

0   100   200   300

Mysteriously, only Brazil had significant microcephaly

Note: Data as of Oct. 20

Source: Pan American Health Organization   THE WASHINGTON POST

If scientists had skepticism re: Brazil's #'s, there'd be less "bewilderment" about absences elsewhere

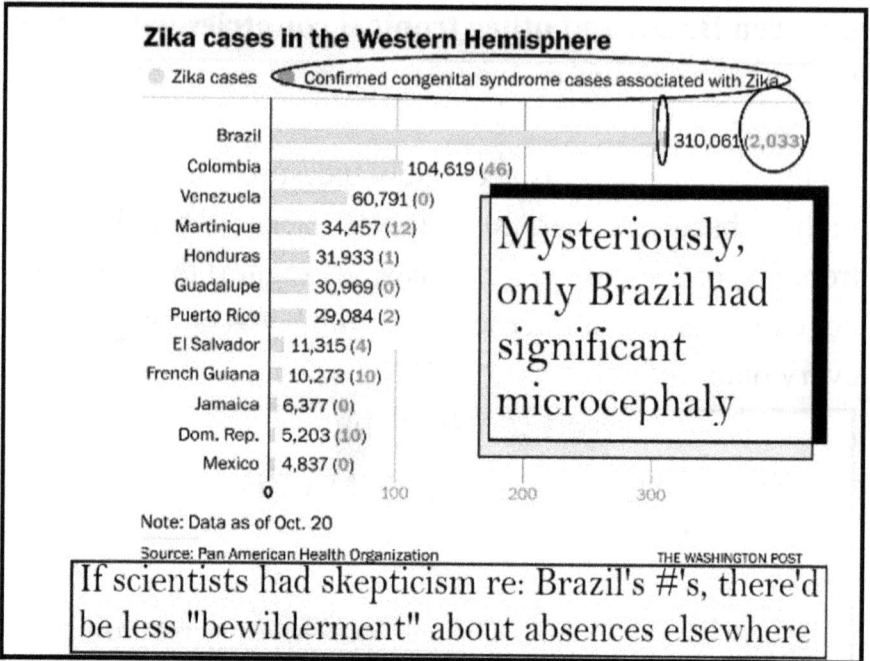

Brazil's immediate neighbor, Colombia (its southern portion sharing the Amazonian tropical rainforest, Zika and Aedes aegypti) didn't show Brazil's Northeast's microcephaly (see below, black circle).

Neighboring Colombia shares Brazil's Aedes aegypti Zika distribution; but only at 1/15 Brazil's microcephaly rate – and 2% of case #s

Microcephaly in Brazil

Microcephaly in Colombia

from: A Possible Link Between Pyriproxyfen and Microcephaly. PLoS Curr. 2017 Nov; Bar-Yam Y.

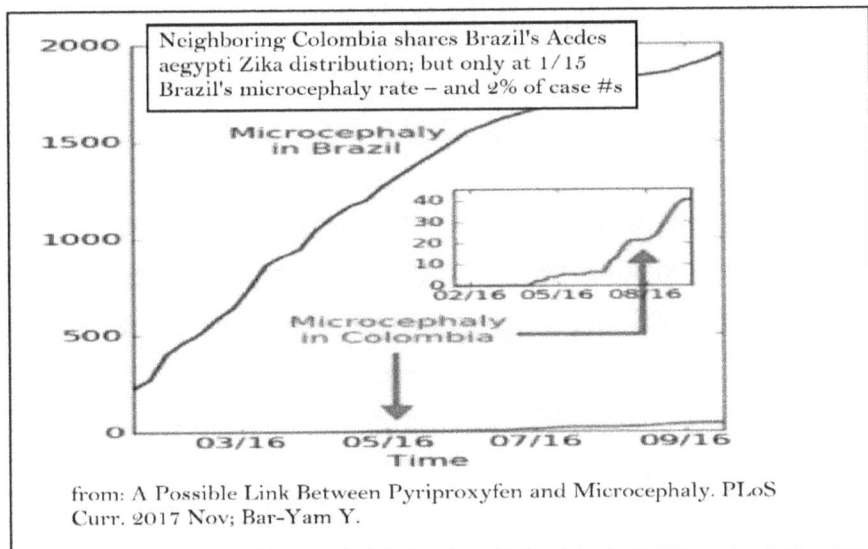

To be fair, neither did Brazil's own state of Amazonas on Colombia's border. And in the following year basically nobody did, not even Brazil, not even Brazil's Northeast. And every year after that, it's quiet.

What is the explanation for this huge differential in microcephaly rates between
- 2015 Brazil's Northeast versus the rest of its states, 2015
- 2015 Brazil versus 2016 Brazil
- 2016 Brazil versus Latin America
- 2015 Brazil versus every other country (including Brazil) every year thereafter

Clearly something unique and un-reproducible happened in Brazil, 2015. For some reason nobody seems to think it could be that the intrepid souls who came up with any or all of the three parts constituting the Zika-microcephaly theory could be wrong.

Brazil's Northeast is not unique geographically. There are parts of South America with similar rainfall, similar temperature, similar topography and indeed all three. It is not unique either in being home to the Aedes aegypti mosquito. So conditions that were operative in Recife, Pernambuco in bringing about this theory's increase in Zika and later microcephaly should have been duplicated in other places, if real. These maps perhaps help prove the point.

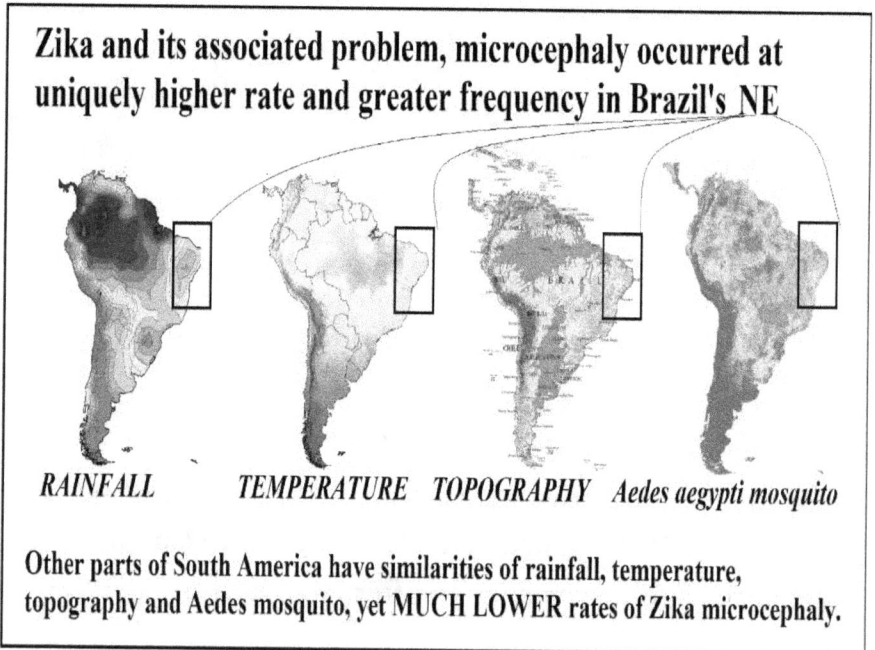

Zika and its associated problem, microcephaly occurred at uniquely higher rate and greater frequency in Brazil's NE

RAINFALL   TEMPERATURE   TOPOGRAPHY   *Aedes aegypti mosquito*

Other parts of South America have similarities of rainfall, temperature, topography and Aedes mosquito, yet MUCH LOWER rates of Zika microcephaly.

Viral Penetration, Rubella versus Zika

The viral triad of "measles, mumps, rubella" is well known by its acronym "MMR", the mandatory childhood vaccination. Each of these three viruses brings a moderate childhood illness parents would just as soon have their kids avoid – but the gravest reason for this shot is girls' obtaining immunity long prior to pregnancy, preventing rubella's dangerous, congenital neurologic -damage: principally the baby's mental retardation and/or deafness' resulting from essentially all first trimester maternal rubella infections.

CONSEQUENCES OF CONFIRMED MATERNAL RUBELLA AT SUCCESSIVE STAGES OF PREGNANCY

ELIZABETH MILLER   JOHN E. CRADOCK-WATSON
THOMAS M. POLLOCK

Over a thousand women with confirmed rubella infection at different stages of pregnancy were followed up prospectively. ... **The frequency of congenital infection after maternal rubella with a rash was more than 80%** during the first 12 weeks of pregnancy.... **Rubella defects occurred in ALL infants infected before the 11th week** (principally congenital heart disease and deafness).... In contrast, none of the 11 infants whose mothers had symptomless rubella was infected.

The Lancet · Saturday 9 October 1982

Through vaccination's success, *"Rubella Syndrome"*[140] has essentially disappeared. For that reason, medical literature on the topic may appear outdated. This 1982 review showed that rubella's rate in inflicting damage is nearly universal. *"Rubella defects occurred in **all** infants infected* *before the 11th week (principally congenital heart disease and deafness) and in 35% of those infected at 13-16 weeks (deafness alone)."*[141]

This is a quite different situation from Zika's, which ***AT WORST*** was originally thought to have a penetration rate of congenital defect less than 10%. Even that low percentage is higher than the ultimate reckonings: with guesses from researchers now hovering in the low single digits of percentage, 2%- 4%. This congenital-damage rate is anywhere from 10 to 50 times lower than other congenitally neurotoxic infective agents'.

Rubella had been the most common of these three, but has disappeared with universal vaccination. The other two are only quite rarely diagnosed, therefore data are hard to come by. Birth defects occur with a 30-40% frequency from maternal infections from either cytomegalovirus[142] and the protozoan parasite, toxoplasma.[143]

For a brief time in the early 1960s, thalidomide was a prescribed sleeping pill. It brought about the sad occurrence of babies without limbs, born to unsuspecting mothers. The pill was immediately withdrawn when the connection was made. Retrospective studies showed that the probability of birth defects was 50% from ingestion of a single pill within the first 30 days of pregnancy – and likely close to 100% for multiple episodes.[144]

To summarize, here are congenital-damage rates from various agents, with first-trimester exposure:

agent	congenital damage rate
Rubella	80-100%
Cytomegalovirus	30-40%
Toxoplasmosis	30-40%
Thalidomide	50-100%
Zika	2-4%
Dengue	0%

Sesame Street's song *"One of these things is not like the other, one of these things just doesn't belong"* is not precisely scientific, but it does serve in this case as a reminder that Zika's unlikelihood in causing congenital damage reflects the unlikelihood that it causes congenital damage. That's a bit of a tongue twister, but hopefully the point is made: if it actually did bring a dangerous process *in utero*, it would likely be perpetrating such damage on a more regular basis. The other infective agent culprits are 10 to 40 times more potent and frequent in doing so. This casts doubt on Zika's being a member of that notorious club. Zika seems closer to safe-pregnancy dengue.

Another difference between Zika and these more reproducibly, long-term documented teratogens is that the others had broader, better established bases of comparison from which to establish the damage rates. ***Overturning Zika*** documented that 2015 Brazil had no prior registry of microcephaly.[145] Although the Drs. van der Linden presumed Recife's microcephaly rate to be higher than baseline, later data reconstructions demonstrated that not to be the case.[146]

Furthermore, even if there had been a registry, the "baseline" would have been higher than in more developed countries. The conditions of extreme poverty, themselves, track with higher microcephaly-rates – so, if and when a comparison would have been made in Recife 2015 there would not have been much to notice in the way of "increase".

This phenomenon was interestingly brought to light in 2018 by Brazilian researchers publishing in PEDIATRICS. They performed *"A Tale of Two Cities"* within two lesser-known towns 1600 miles apart, with few other similarities aside from both being in the nation of Brazil: inland versus oceanside; South versus North. [147]

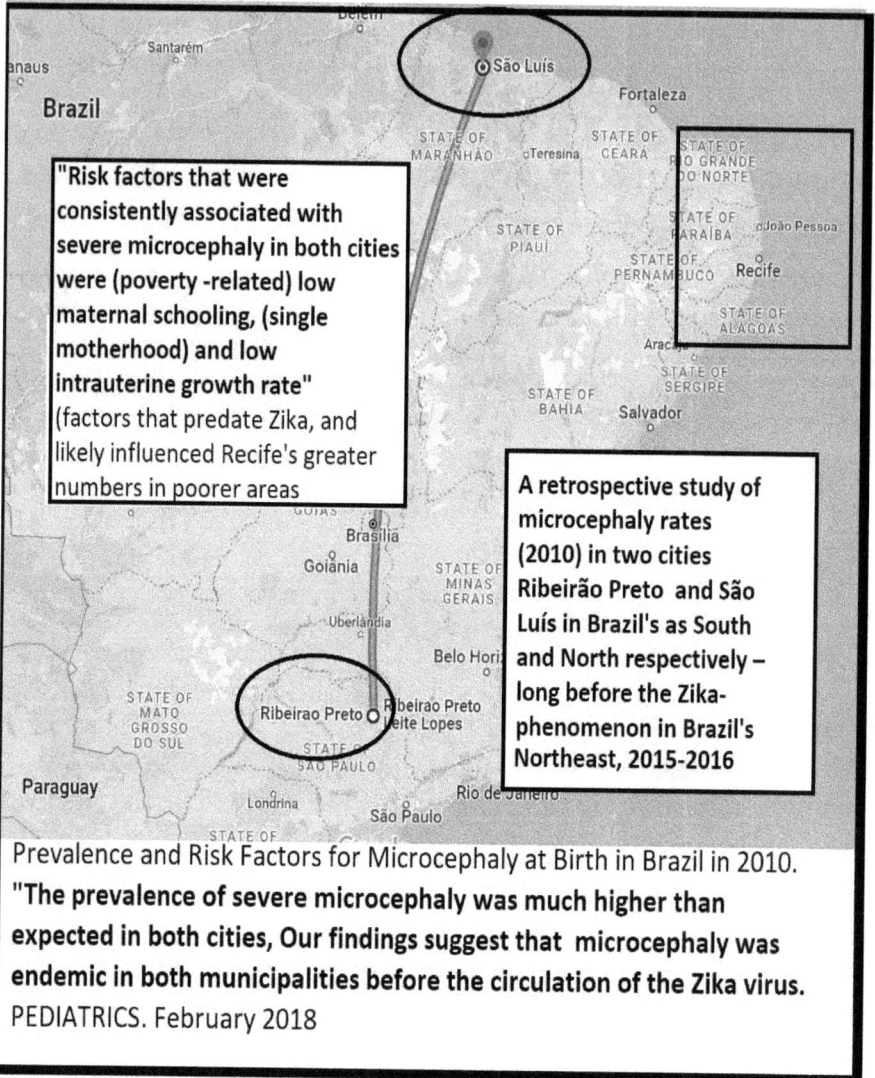

"Risk factors that were consistently associated with severe microcephaly in both cities were (poverty -related) low maternal schooling, (single motherhood) and low intrauterine growth rate" (factors that predate Zika, and likely influenced Recife's greater numbers in poorer areas

A retrospective study of microcephaly rates (2010) in two cities Ribeirão Preto and São Luís in Brazil's as South and North respectively – long before the Zika-phenomenon in Brazil's Northeast, 2015-2016

Prevalence and Risk Factors for Microcephaly at Birth in Brazil in 2010. **"The prevalence of severe microcephaly was much higher than expected in both cities, Our findings suggest that microcephaly was endemic in both municipalities before the circulation of the Zika virus.** PEDIATRICS. February 2018

This study would seem to imply that the real "virus" bringing on higher rates of microcephaly in Brazil (and likely elsewhere around the world once looked at more closely) is poverty – not directly, but through the various risk factors that go along with it. We began an earlier chapter with this table, showing:

<u>**These underlined microcephaly associations**</u> occur more frequently in poverty-conditions (such as Northeast Brazil's).

> <u>**Disruptive injuries**</u>
> <u>**Infections: "TORCHES" (toxoplasmosis, rubella, cytomegalovirus, herpes varicella, syphilis) and HIV**</u>
> <u>**Poorly controlled maternal diabetes**</u>
> <u>**Deprivation**</u>
> <u>**Maternal hypothyroidism**</u>
> <u>**Maternal folate deficiency**</u>
> <u>**Maternal malnutrition**</u>
> <u>**Alcohol-overuse**</u>
> Teratogens: hydantoin, radiation
> Maternal phenylketonuria
> Placental insufficiency
> Death of a monozygous twin
> Ischemic or Hemorrhagic stroke

Plenty of individuals in poverty don't have these issues, but there are correlations between poverty, and drug- and alcohol- use; domestic violence; spottier nutrition; proximity to toxins; circulating germs via mosquitoes; and less hygienic living conditions – all of which can indirectly inhibit growth during pregnancy.

Poverty also tracks with prematurity and <u>in Brazil, smaller stature</u> – both of which can bring smaller head circumference (**SHC**) at birth. **SHC** can be confused with microcephaly (if proportionality to the body's overall smaller size is not checked simultaneously). Poverty may very well have been the chief generator of cases of microcephaly in Recife 2015 as we see in this very <u>next chapter</u> (independent of whether actually there was a true "microcephaly epidemic").

A closer look at Brazil's Zika-microcephaly epicenter, Recife Brazil

This next chart displays cases of microcephaly in Recife, plotted geographically against an overlay of wealth disparity between the city's neighborhoods.[148]

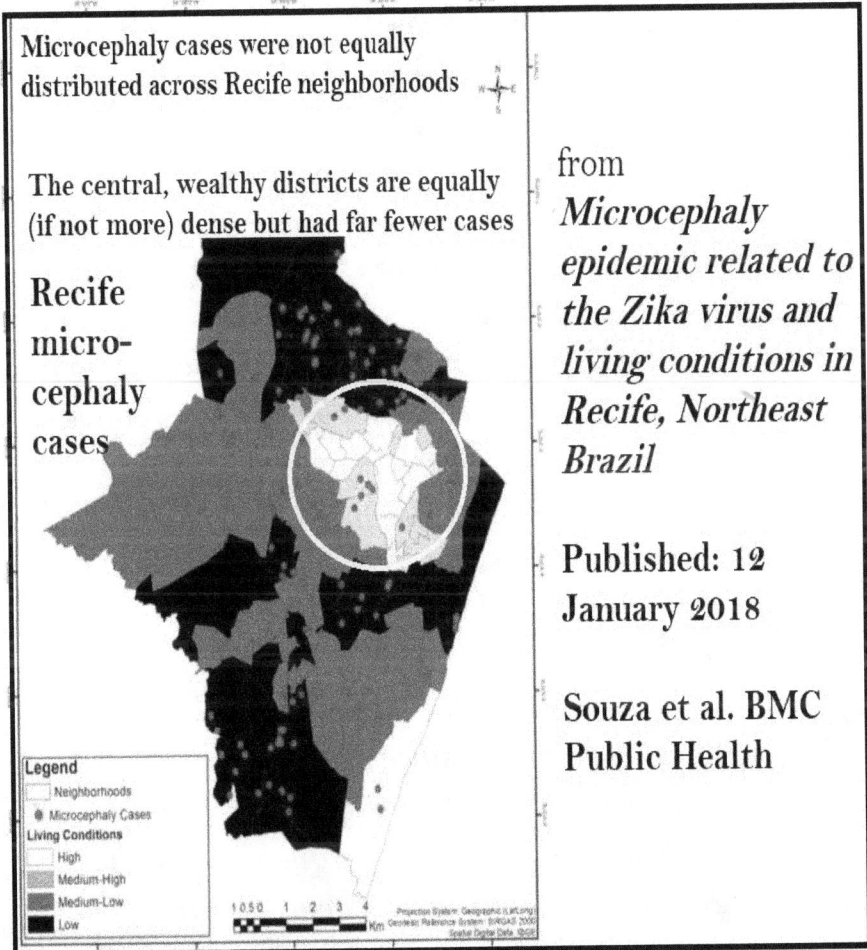

Microcephaly cases were not equally distributed across Recife neighborhoods

The central, wealthy districts are equally (if not more) dense but had far fewer cases

Recife micro-cephaly cases

from *Microcephaly epidemic related to the Zika virus and living conditions in Recife, Northeast Brazil*

Published: 12 January 2018

Souza et al. BMC Public Health

Legend
Neighborhoods
Microcephaly Cases
Living Conditions
High
Medium-High
Medium-Low
Low

Each dot represents a fixed number of microcephaly cases within Recife, 2015, concentrating in areas of poverty.

Central Recife is more densely populated (with high-rises) yet there were fewer microcephaly-cases, prior to 2016's extraordinary precautionary measures to avoid mosquitoes The same holds true for the densely populated, wealthy beach district, Boa Viagem (southeast, i. e. lower right).

Americans may be used to its suburbs' being wealthier. Brazilian cities reverse this pattern. The wealthy concentrate in the city center while the poor reside in the surrounding areas, possibly the historical result of the abolition of slavery in 1888.[149] Freed slaves were pushed away from the wealthy city centers and forced to reside in the outskirts.[150]

> "Since Recife is a capital city, it isn't uncommon for *camponêses* to arrive seeking better living conditions and end failing this quest. Being left with literally nothing, many build makeshift houses at the edge of dirty canals and make a living in any way that is available to them; this is the "extreme situation". Some mix into the shantytowns, but to do so need some money. The favelas have an interesting "survivalist contrast" since they are considered "forgotten" or marginalized spaces. People act as "gatos", stealing electricity through wire-grafts and digging holes in the water system. Effectively, the cost is covered by the rest of the city, who end up having to pay more. There is a wide range of economic contrast in Recife, from extremely rich to absolutely miserable. Some people literally live

"under the bridge". Open sewage is a common issue in Northeastern Brazil."[151].[152] [153]

A 2016 Brazilian study showed only 38% of Recife's urban population was connected to the sewerage system.[154] Most of the poor can't afford air conditioning. Tropical weather, open windows without netting, plentiful sites of stagnantly pooled water contribute to these areas' having more exposure to mosquitoes than their wealthier counterparts; however, before jumping to Zika conclusions, be aware that multiple other congenital microcephaly co-factors associate with poverty.

Another crucial factor to consider is the differing ethnic mix of more *mestizo* and *"Indio"* in Northeast Brazil vs. taller, more ethnically European in the South / Southeast (Brasília, Rio, São Paulo) [155] where (as the locus of Brazil's power, prestige and wealth) the erstwhile head-measurement standards charts had (in part) originated. [156]

Distribution by color or race by region			
region	White	Black	Multiracial
North	23.2	6.5	67.2
Northeast	29.2	9.4	59.8
Center-West	41.5	6.6	49.4
Brazil	47.5	7.5	43.4
Southeast	54.9	7.8	36
South	78.3	4	16.7

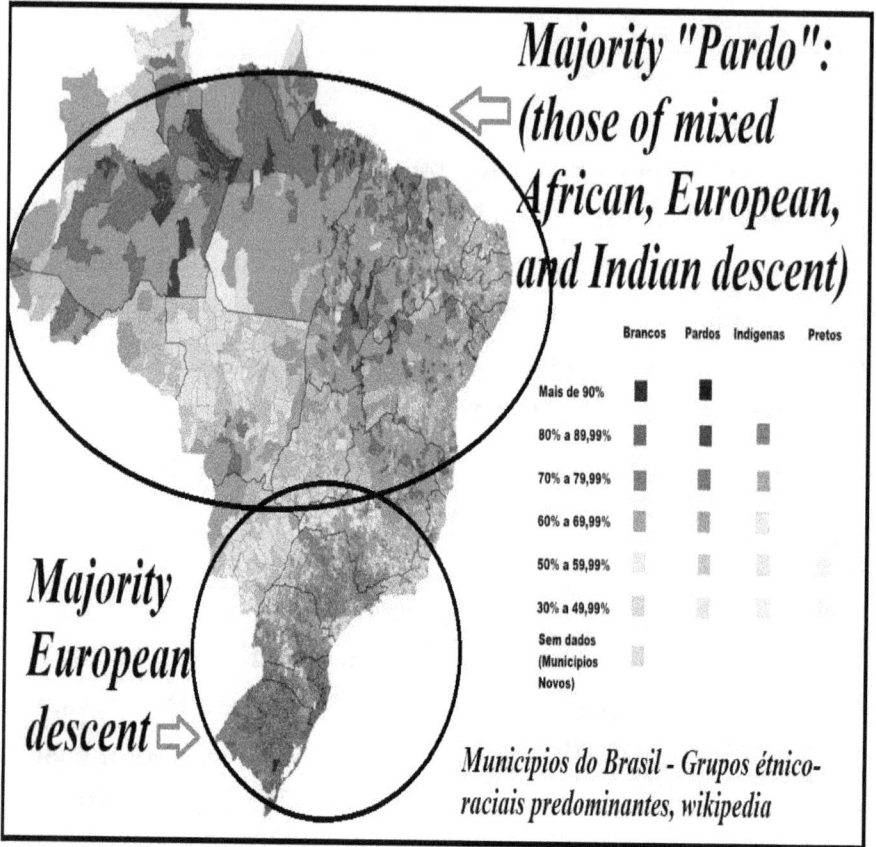

Majority "Pardo": (those of mixed African, European, and Indian descent)

Majority European descent

Municípios do Brasil - Grupos étnico-raciais predominantes, wikipedia

Roughly, height increases from Indio, to mestizo, to white[157]. Brazil's South and Southeast historically have been taller than the regions of the Zika/microcephaly outbreak.

Regional differences in height, Brazil

'European' South ~5 cm taller than Northeast

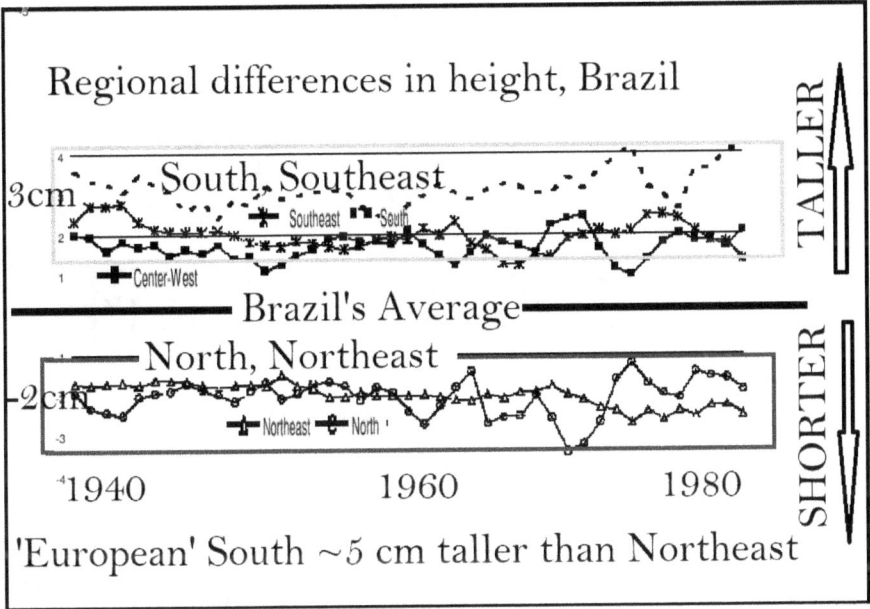

The causation is likely a mix of innate genetic differences with the growth-stunting aspect of poorer nutrition[158]. There is an additional factor of the height-boost from intermarriage's hybridization, but determining its rate-differential between ethnic groups is beyond this paper's focus. In any event, smaller individuals breed smaller babies. Additionally poverty-circumstances track with more prematurity, itself a force for smaller head circumference at birth.

Head Circumference (Hadlock1984)

Babies born prematurely if mistaken as full-term can be misinterpreted as having had abnormal head size

+ 2SD

Mean

- 2SD

HC (Centimeters)

Premature birth

Normal gestational age at delivery

misinterpretation possibly as microcephaly?

Gestational age in weeks

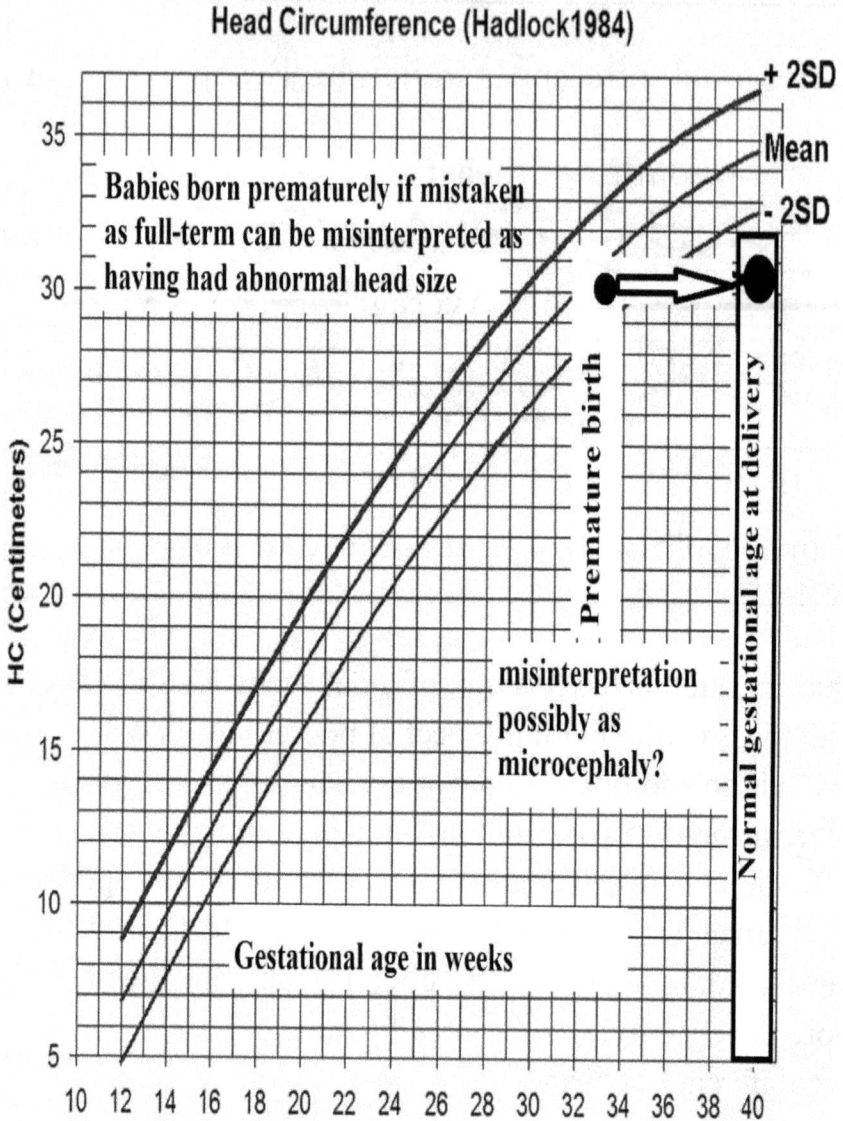

Standard head-size measurements derived from taller populations with overall better nutrition and less prematurity may not have been the best choice to establish (de novo) whether in fact there was increased microcephaly

in these poorer parts of Recife. Children born and raised in wealthier areas would be expected to grow bigger and taller—head size included. If this was indeed the case, there's a real possibility that microcephaly in these poorer areas might have been over-diagnosed in the rush to judgment. All of these pronouncements should have come with HUGE DISCLAIMERS that there had been no prior year comparisons in the same-ethnicity, same demographic of Recife-poverty.

[As mentioned in the previous chapter] In 2018, a couple of years after the fact, Prof. of Epidemiology Antônio da Silva attempted to (re-)create a context for Recife's presumed microcephaly surge, 2015. He took data from 2010 measured in two Brazilian cities, one North, one South. Using two different background standardized charts, he found higher than predicted rates in both cities:

> *"Our findings suggest that **microcephaly was endemic** in both municipalities **before** (!) the circulation of the Zika virus."*[159]

This is stunning! Let's step back for a moment and parse what this means. First of all the non-contextual 2015 Zika-microcephaly increase seems even more clearly to have been confusion wrapped up in desperation and covered with haste. With background microcephaly numbers, 2010, much higher than expected, the 2015 data represents no increase at all!

In the very same issue of Pediatrics, American epidemiologists Elizabeth Dufort and Jennifer White discuss why Prof. da Silva's re-creation was necessary:

> As the world learned of the emerging Zika virus outbreak, ... questions quickly arose regarding the dramatic increase in reported microcephaly.... There were 2 overarching questions at that time:
>
> - How large was the increase in the incidence of microcephaly in Brazil?
> - and, how much of it could be attributed to Zika virus?

These American professors, echoing some of the concerns of this paper, wonder)

> - **Whether increased awareness may have led to increased reporting.**
> - If initial assessments were further compounded by comparison with a pre-Zika baseline that relied on a passive surveillance system.
> - Whether "the lack of an internationally accepted, standardized approach to the measurement and definition of microcephaly, (made) comparisons of rates across regions and time periods difficult.[160]

This chart (very similar to the one shown previously) points out that microcephaly claims not only didn't match

with Aedes aegypti distribution,[161] but they didn't match with Brazil's population centers. The strongest determinant of claims would seem to be proximity to Recife, the site of the original "microcephaly epidemic" claim. [162] Notably that's not specifically the area where "mild dengue" was determined as "Zika".

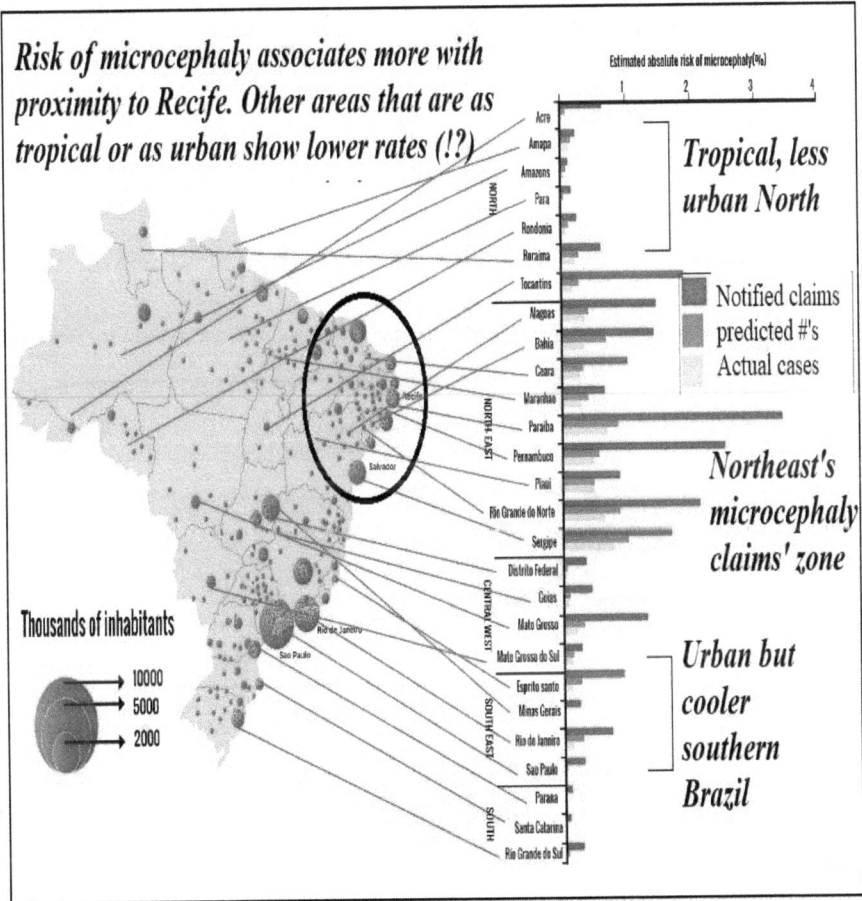

Risk of microcephaly associates more with proximity to Recife. Other areas that are as tropical or as urban show lower rates (!?)

The highest rates occurred in Paraiba and Pernambuco— the areas where people were talking about microcephaly the earliest and the most.[163] Humans act and react emotionally: the more people talk about microcephaly, the more it makes the news. That cycles into even greater worry about it and more reported cases. This happens regardless of whether those worries are medically justified, or not.

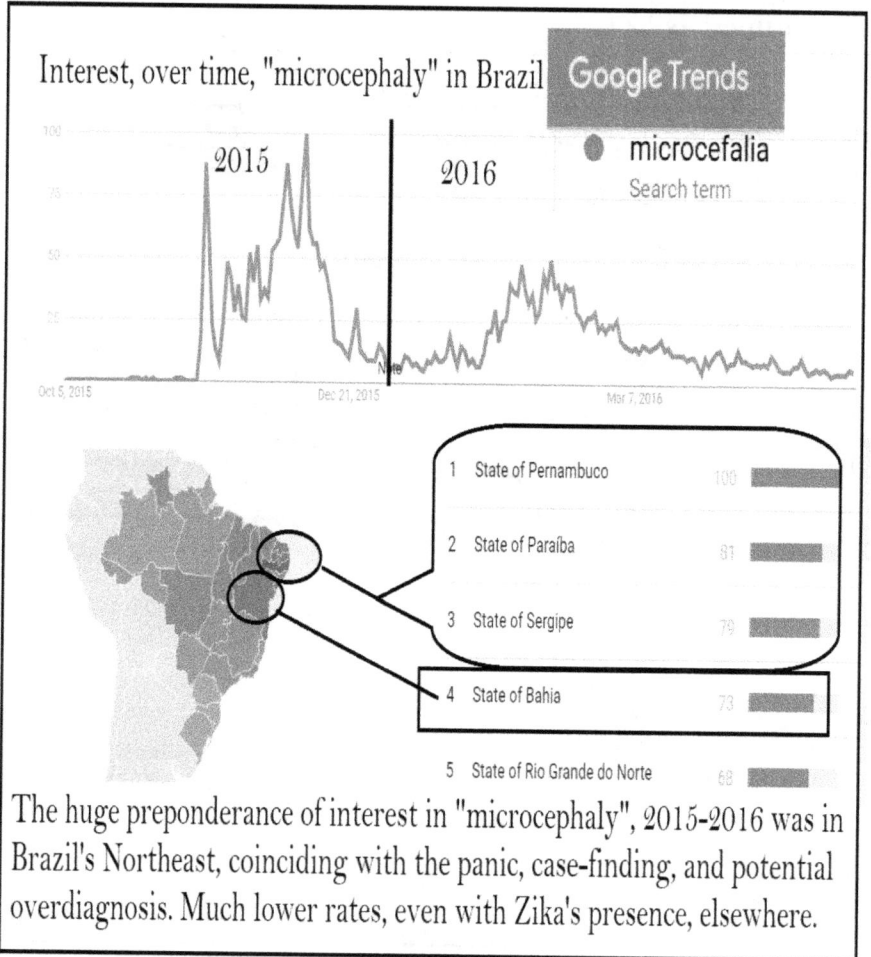

Interest, over time, "microcephaly" in Brazil — Google Trends

microcefalia
Search term

2015   2016

1	State of Pernambuco	100
2	State of Paraíba	81
3	State of Sergipe	79
4	State of Bahia	73
5	State of Rio Grande do Norte	68

The huge preponderance of interest in "microcephaly", 2015-2016 was in Brazil's Northeast, coinciding with the panic, case-finding, and potential overdiagnosis. Much lower rates, even with Zika's presence, elsewhere.

Hypothesis Testing: When Results Fail Beliefs

"As night follows day" another year eventually rolls around. After the perceived horrors of 2015's microcephaly, more and worse were expected for 2016. Public health specialists believed Zika was twice as prevalent. Predictions were made for a second microcephaly epidemic (not quite twice the size, because of counterbalancing anti-mosquito measures: draining swampy areas; spraying insecticide; removing trash; staying indoors, using netting and applying insect repellent, particularly during pregnancy.

What nobody expected was that microcephaly would essentially disappear. But that is what happened.[164] No doubt, the drop got some help from Brazil's newly stricter microcephaly criteria's eliminating the prior year's over-diagnosed borderline-normals– along with more accurate data-collection.

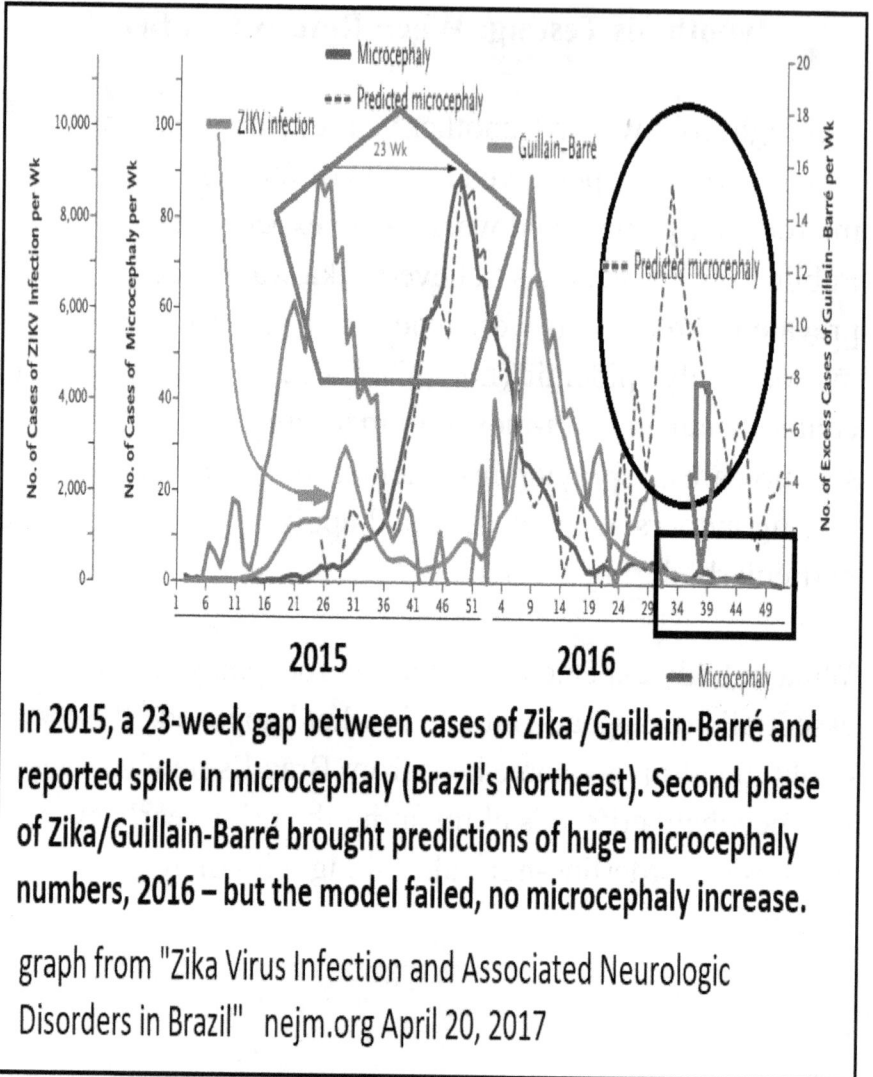

In 2015, a 23-week gap between cases of Zika /Guillain-Barré and reported spike in microcephaly (Brazil's Northeast). Second phase of Zika/Guillain-Barré brought predictions of huge microcephaly numbers, 2016 – but the model failed, no microcephaly increase.

graph from "Zika Virus Infection and Associated Neurologic Disorders in Brazil" nejm.org April 20, 2017

But another thing that helped get the numbers down? Well, ... *the Zika/microcephaly hypothesis' having been wrong in the first place*!? Statistically, it doesn't appear that Zika and microcephaly have anything to do with each other. Naturally, humans' being humans (and with researchers' never wanting to quash their own funding's rationale), this was not the conclusion drawn. Rather, new theories arose.

Before getting to those, let's look at another retrospective analysis, "*A microcephaly case-control study in an area of high dengue endemicity in Brazil*"[165] submitted in 2019, by researchers associated with Brazil's Microcephaly Epidemic Research Group (MERG) focusing on the Dr. Vanessa van der Linden's Recife "epicenter" of Zika/microcephaly in 2016, one year after the original claims. The differences were these:

- This study took place in the absence of panic.
- There was no flooding of regionally panicked moms' converging on one hotspot of diagnosis.
- These moms were actually laboratory tested for Zika (and as well for dengue)
- Microcephalic-babies' moms were compared to a control group (not just singled out and individually contacted as originally done by Dr. Carlos Brito, 2015)
- They used a single microcephaly standard (although it was essentially the wrong one, not the WHO's intergrowth chart, stricter in its criteria; rather an across-the-board lowest 2.5% of all babies – which necessarily will include many small but normal babies. This is an error, but at least it was consistent, as opposed to 2015's lack of a consistent standard)
- The results were compiled and compared before being broadcast.

- The researchers acknowledged how difficult *and potentially nearly-impossible* it is to differentiate Zika from dengue. Their conclusion:
  - *"**Laboratory confirmation of ZIKV infection during pregnancy is challenging due to cross-reactivity with other flaviviruses, especially dengue.** The neutralization test, which is the gold standard to discriminate between these viruses, **is time-consuming, performed in few laboratories and does not define the time** when the infection occurred."*

One similarity was that these small headed babies were never matched or correlated with actual health issues, e.g. mental retardation, reminding us that the original pandemic was not fully associated with cognitive impairment , but rather ONLY a size metric.

Between these two groups (of 89 mothers of microcephalics and 173 moms of normal-sized babies), *there was no real difference as far as infective Zika or dengue exposure*.

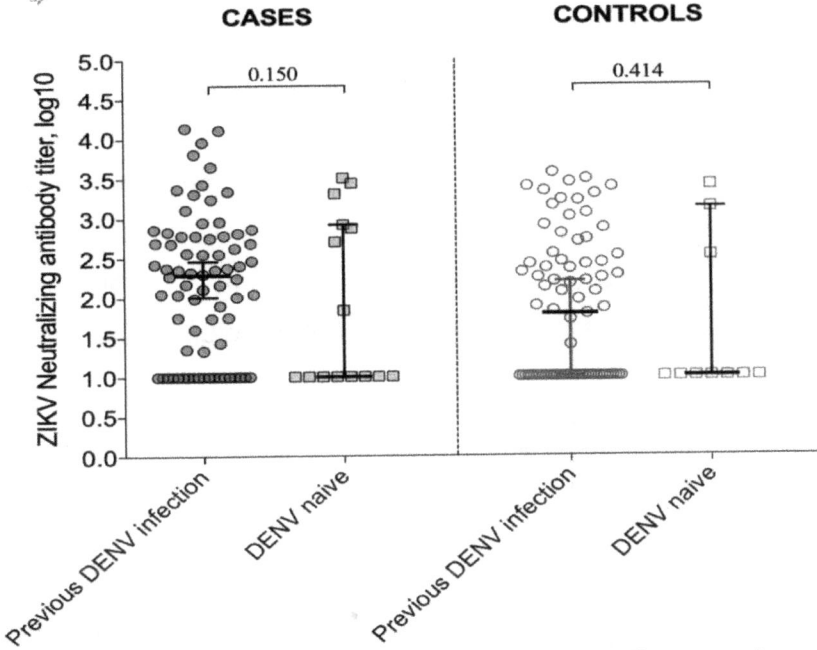

ZIKV-specific NAbs titers (log10) among mothers according to previous DENV exposure, i.e. immunity to Zika overlaps with dengue, & vice versa

Only one in 11 moms from either group was fully immunologically naïve to the flaviviruses genus (i.e., never had had either Zika or dengue). The remaining background percentages amongst mothers for Zika and dengue combination exposures were essentially the same for the two groups. Fewer than 10% had been exposed *only* to Zika and never to dengue, with no difference in this aspect between two sets of mothers. You can see from the diagram that the moms of microcephalic cases had very similar Zika- AND dengue- exposure rates to the (normal) control set. Those exposures seem immaterial to microcephaly status.

Two small side notes:

- Does this panorama and panoply of multiple dengue infections mixed in with the occasional Zika look as if Zika had taken over for dengue amidst the Aedes aegypti mosquito population? Based on how the mothers present after presently having been bitten by those very same mosquitoes, the answer is *"no, not really"*.

- There's only a tiny increased likelihood of Flavivirus infection in the microcephalics' moms. This might be meaningless or it might imply more poverty-generated other causes' conflating the data.

This study's results essentially deny a Zika-microcephaly connection. Do the researchers present it that way? Do they voice skepticism of the underlying hypothesis? *"No"*, and *"no"*; instead we get the standard (kicking the can down the road) reply, *"more research needs to be done"*, etc.:

> *"Considering the relatively low frequency of markers of recent ZIKV exposure at delivery, screening for ZIKV immune status should be performed in the early stage and throughout pregnancy to monitor congenital ZIKV syndrome in endemic areas. Innovative laboratorial diagnostic approaches for ZIKV and DENV infections are urgently needed for the guidance of clinical practice and public health purposes."*

Here are some further cases of those very heavily invested, personally, financially, or reputationally in the Zika-microcephaly causation hypothesis' not being able to "fess up" that it might not actually have been the case.

- In 2017 Christopher Dye of the WHO admitted, "*We apparently saw a lot of cases of Zika virus in 2016.* ***But there was no microcephaly. The difference (between 2015 and 2016) is spectacular.***[166] *First off,* *"health officials could have vastly overestimated the number of Zika cases in Brazil. So chikungunya can easily be mistaken for Zika,"* [to which NPR, to its credit, blithely responds: **"But chikungunya doesn't cause microcephaly."**]

- Dr. Albert Ko Yale's School of Public Health chair of Epidemiology, went in a different direction, mildly rebuking Dye:

  o *"Misdiagnosis is a reasonable hypothesis. But it's not clear that this explanation accounts for the whole story. Zika might not be working alone. Maybe another infection combines with Zika to make the disease worse and increase the risk of birth defects. But many groups are concerned about the exposure to dengue in Brazil. So another hypothesis is that prior exposure to dengue may actually enhance or promote the risk of birth defects from Zika,"*

Ko in 2017 began working on epidemiological studies in northeast Brazil to see whether that is the case -- presumably having been granted funds for that purpose and then not having the theory pan out (or else we would have heard of it by now). The 2016 Recife microcephaly data presented here seems to refute his theory: dengue frequency is basically the same in normals' as in microcephalics' moms. In any event, NPR concludes its article, *"If (Dr. Albert Ko's) dengue theory turns out to be true* ["and even if it isn't", NPR should have added], ***it could mean the global threat of Zika for pregnant women is less dire than scientists originally thought."***

Insofar as the topic is revisited occasionally in the popular press, it is to keep the fear up, not to give people relief even though the original Zika-microcephaly theory now seems more wild conjecture than actual science. Did **NPR** follow that hopeful hint anytime later in these subsequent years of Zika-quiet to suggest the end of Zika's threat for pregnant women? *"No"*, and The New York Times (**NYT**) hasn't either, writing in 2019:

- *"Brazil remains vulnerable"*
- ***"The next outbreak is not a matter of if, but when"***
  -- Dr. Ernesto Marques
- ***"Zika has completely fallen off the radar, but*** *the lack of media attention doesn't mean it's disappeared. In some ways,* ***the situation is*** *a bit* ***more dangerous*** *because people aren't aware of it."*
  -- Dr. Karin Nielson

In 2022, the New York Times returned with, *"The Forgotten Virus: Zika Families and Researchers Struggle for Support"* -- presenting researchers still unwilling to voice any doubt in the original Zika-microcephaly theory, despite three more years of an absence of the connection between the two, worldwide. Excuses continue to abound:

- *"It's still as urgent as it was back in 2015: We still need better ways to diagnose Zika infection. My suspicion is there is transmission, but it's not hitting the books, it's not being detected."* Epidemiology Chair at Yale School of Public Health, Dr. Albert Ko
- *"You see 97 percent of the cases are in lower socioeconomic classes, and only 3 percent in medium and high. It could be things associated with poverty in some way facilitate the virus to cross the placenta."* Dr. Ernesto Marques
- *"Are Indians and Thais less susceptible, or are we just not detecting it? Is congenital Zika syndrome being misdiagnosed as something like toxoplasmosis?"* Dr. Albert Ko

Most of these experts are "half right" in the sense that Zika is indeed currently "less dire a threat", but they are "half wrong" for this more important reason: that is it may never have been a threat, in the first place:

- It was never diagnosed in 2015 in a single individual, pregnant or otherwise.
- It was confused with pre-existing dengue in the field, the clinic, and laboratory
- If it did (or does) exist in Brazil it's probably a subset of dengue
- Neither of which cause microcephaly (dengue's demonstrably proven in this regard, having been around far longer)
- Once all the criteria were standardized in Brazil and physicians put under closer scrutiny for data collection, microcephaly was shown to be at "normal" background levels: minimal and not pandemic in 2016 or even in the prior "pandemic" year.

In logic, a type II error (false-negative) occurs if the investigator fails to reject a null hypothesis that is actually false in the population. It may be that we are seeing a fair number of these, still, in the Zika-field.

Reality

	True	False
True	Correct ☺	Type 1 error False Positive
False	Type 2 error False Negative	Correct ☺

Measured or Perceived

So, the central prediction of the theory of a Zika epidemic failed to pan out. Shortly thereafter, Zika faded from both the national and the international consciousness. Dengue fever remains a problem in Brazil. More recently, COVID-19 has preoccupied the country. No one talks much about Zika or the panic it caused.

None of this is meant to minimize the suffering of parents of children born with microcephaly. Microcephaly is a tragedy regardless of its cause. However, if we as a global society are going to deal with viruses and the epidemics and pandemics they have the potential to generate, we must learn to do so constructively. We cannot fight future pandemics by diverting scarce financial and research resources to problems that don't actually exist. Our knowledge of viruses and their effects on human populations must be based in scientific fact, not media sensationalism. If there really was a Zika-caused microcephaly epidemic in Brazil in 2016, let the science show it. The current general consensus on the matter lacks adequate scientific backing.

In summary, a previously completely-unknown and pathologically-benign Zika was never actually measured in a single patient at the time it was theorized and then announced as existing in Brazil (for the first time) -- whilst indistinguishable from its biologic cousin: long-standing, locally-endemic dengue-fever. Separately, Recife's microcephaly pronouncements relied on incorrectly-templated standards aggrandizing an unfortunate set of births which were never proved or statistically documented to represent an actual increase over perennial and sporadic baseline, random occurrence. These two individually unverified and unverifiable, unrelated occurrences (Zika and microcephaly) were improbably brought together as one's having caused the other. Whereupon the media-floodgates opened, bringing a bystander-effect of more cases' being announced under a panic of two simultaneous epidemics. It took quite some time retroactively and retrospectively to piece together documentation that each leg upon which the Zika-microcephaly connection stood not had not solidly been constructed. That it has remained

upright for this long implies its having been propped up, externally -- while funded perpetually.

Never Let a Crisis Go to Waste

"You never let a serious crisis go to waste. And what I mean by that is it's an opportunity to do things you think you could not do before."[167] [168] Rahm Emanuel, 2008

Your Bodies, Our Choice; Zika's Key Unlocking Abortion

For many, Zika presented that opportunity to right perceived wrongs. Brazil, like every other country, has its internal divisions, rivalries, jealousies -- and factions of geography, class, race, and education. The Zika-microcephaly phenomenon embodied a lever by which Brazil's poor Northeast could gain the attention of, and finances from, Brazil's power-center trio of Rio de Janeiro, São Paulo, and Brasília.

Pro-abortion activists found opportunity amidst Zika misfortune to undo Catholic Brazil's restrictions– at first temporarily using emergency injunctions; as well as aiming at permanent reversals.

The Guardian	**THE WALL STREET JOURNAL.**
Zika emergency pushes women to challenge Brazil's abortion law	**Brazil's Attorney General Asks High Court to Allow Abortions for Women With Zika**
Women's groups are set to challenge the law in the hope of making termination possible for women at risk of delivering a baby born with Zika-related defects	The request has sparked objections in this socially conservative country
July 19, 2016	**September 8, 2016**

"The fears over the Zika virus are giving us a rare opening to challenge the religious fundamentalists who put the lives of thousands of women at risk in Brazil each year to maintain laws belonging in the dark ages." Silvia Camurça, a director of SOS Corpo, a feminist group in Recife [169]

Anis Institute filed with Brazil's Supreme Court similar demands, incorporating class and racial division within pro-abortion argumentation:

> *"The (Zika-) epidemic mirrors the social inequality of Brazilian society, concentrated among young, poor, black and brown women. They live in substandard, crowded housing in neighborhoods ... most likely to contract Zika. The government must finally give women ... safe and legal abortion."[170]*
> Debora Diniz, attorney, anthropologist, Anis.

When Brazilian Judge Jesseir Coelho de Alcântara publicly stated abortion should be allowed in microcephaly cases, local activists cited California's 1967 Therapeutic Abortion

Act– which brought some of the first legal abortions in the United States – to avoid (pre-vaccine era's) Rubella's birth defects (which occurred with [at least] <u>an order of magnitude greater frequency than the worst-case conjectures regarding Zika's</u>).[171]

---

242   THE HASTINGS LAW JOURNAL   November, 1967]

THE CALIFORNIA THERAPEUTIC ABORTION ACT: AN ANALYSIS

Permitting abortions for eugenic reasons constitutes the most controversial ground for terminating a pregnancy because an abortion on these grounds involves a certain degree of medical speculation whether or not the child will be born defective.

However, certain conditions occurring during pregnancy are known to cause a relatively high percentage of serious birth defects. Examples of such conditions are:

(1) **disease of the mother during pregnancy, such as rubella;**

(2) harmful drugs taken by the mother during pregnancy, such as Thalidomide;

(3) irradiation to the pelvic region before pregnancy is discovered; and (

4) evidence of serious genetic defects.

The risk that any child will be born with serious defects as a result of these factors varies between 20 and 60 percent

---

Support was offered from academic circles, as here (somewhat ghoulishly), by Prof. Caitlin Killian, putting the purported microcephalic babies to social use:

> *"The **silver lining of Zika** may be that an illness becomes a catalyst for social change by leading*

governments to reevaluate policies, in this case reproductive rights laws."[172]

The Yale Global Health Review weighed in:

"For the millions of women affected by Zika in Latin America, reproductive health education and comprehensive support are key to helping them escape poverty, realize their rights, and live lives of dignity. This is not just a fight for public health, but also one for women's rights and social justice."[173]

Brazil was seen as "Ground Zero" in the battle to overturn not just theirs but other potentially Zika-affected tropical, Catholic countries' pro-life prohibition statutes.[174]

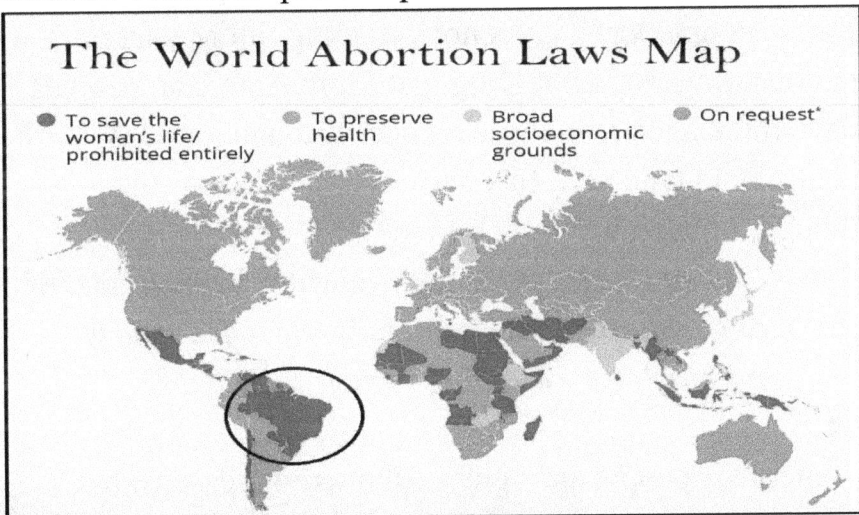

The World Abortion Laws Map

● To save the woman's life/ prohibited entirely   ● To preserve health   ● Broad socioeconomic grounds   ● On request*

Zika's chronicler, **NYT**'s Donald McNeil, covered Latin American nations' calls for a **"zero-child-policy"** exceeding in severity the one Mao conceived (no pun). In reviewing his book, the similarly accomplished science writer Laurie Garrett (in other ways his political ally, fully aligned with his *climate-change brings "emerging threat"* *viruses* -narrative[175]) finds Mr. McNeil's sympathies for this stance *"unnerving"*:

> *"Mr. McNeil favors a universal pregnancy avoidance plan for countries in the grip of Zika, and denounces virtually all sympathy for women facing family planning decisions as* (a version of) *"patronizing" feminism.'[176]*

Ms. Garrett might have gone too far in this criticism. Mr. McNeil -- merely by not condoning certain feminists' giving Latin American women a pass in birth decisions, e.g. "*Zika Virus and the Hypocrisy of Telling Women to Delay Pregnancy*"[177] -- is not yet going full Maoist on childbearing. He's implying that Latin American couples have more agency in those choices through available birth control and type of sexual activity. Make no mistake though, McNeil wants a child moratorium.

> *"When I heard it, I began stewing.* **Women needed to avoid pregnancy somehow. Because clearly nothing else was going to save their babies.**"

Sometimes people are right for the wrong reasons, but it seems as if all parties to this discussion are wrong for the wrong reasons, some of which involve not investigating or evaluating the various unlikelihoods of the Zika-microcephaly theory to begin with. ***Overturning Zika*** proposes that ceasing childbirth is not the proper manner for families to ***"save their babies"***.

Tay-Sachs disease might be a reasonable lens through which to look at this problem. Babies born with this problem usually don't live past four years and have severe mental retardation and neurologic issues. Amongst Ashkenazi Jews, this has, for generations, represented one out of 3000 births. It's only recently that there has been some genetic testing to determine which families and babies will be affected. Throughout all of history, up until this very recent technological moment, should all Ashkenazi Jews have averted all childbirth in order to ***"save their babies"?*** That sounds like a fairly dismal and self-defeating approach to life.

Or let's take the case of one excruciatingly sad step forward. With the Nazis in power in the 1930s, should these same Jews (in Europe) have stopped bearing children entirely? The mortality rate for those children, normal almost entirely, was ultimately 80%. The right answer in both of these cases is not to stop living, but to figure out a way to combat the problem. In the case of Tay-Sachs, it's genetic counseling and chromosomal testing; for Nazi-ism it's fight or flight; for microcephaly, it would be Zika testing and fetal monitoring (if the theory becomes scientifically proven).

This underlines, even more, the scientific and journalistic "malpractice" that occurred with the Zika-microcephaly pandemic's hysteria and exaggeration. People were forced to make life-altering decisions based on claims that had not been verified in almost all aspects. "Microcephaly" encompasses a large variety of syndromes, as is obvious in retrospect (when the number of verified claim cases shrank to only a couple percent of the original number).

Donald McNeil's argument and urgency for zero childbirth in Latin America, indefinitely, are based on avoiding (knowingly) bringing into existence (potentially) some hundreds of "extra" microcephalic infants (many with mental retardation). Is it correct though to attempt to make this *"save"* at the expense of losing hundreds of thousands of completely unaffected children– those never getting the chance to live? These questions parsed as "public health", also involve ethics, morality, religion, values and family. The Catholic Church receives a fair amount of scorn within Mr. McNeil's work and his adversaries', as well – but it's probably fair to say that the Church has more experience in the religion and values business than experts using shaky data to circumvent morality discussions.

A side curiosity, thought experiment, would be if Zika were carried by Aedes albopictus mosquito as well and affected all the temperate regions of earth in addition to the tropical. Would Mr. McNeil have recommended a worldwide moratorium on childbearing, foregoing his own future grandchildren, locally?

Cheerleading "Team-Zika"

When, invariably, not enough was accomplished to rectify perceived social ills during Brazil's 10-month Zika state of emergency, some groups lamented its end. In 2017, Human Rights Watch tossed this grenade:

> *"Brazil has not addressed long standing human rights problems that allowed the Zika outbreak to escalate, leaving the **population vulnerable to future outbreaks** and other serious public health risks."[178]*

Since then, there have been no further outbreaks, nor apologies.

These movements and sensibilities alone might explain the quick media-uplifting of what was otherwise an unsubstantiated, slight increase in a still very rare microcephaly syndrome; however, there was an additional "cheerleading" aspect: rooting for "the little guy". Except when you need one of their products, nobody likes *"Big Oil"*, *"Big Pharma"*, *"Big Banks"*, *"Big Business"*[179]– and potentially in Brazil, 2015 *"Big Medicine"*, the agglomeration of government-influenced medical and research institutions.

The press collaborated in avoiding the skeptical academic review that might have occurred if done in proper order. Instead the Zika-microcephaly story "jumped the queue."

"Major announcements were made to journalists before peer-reviewed publications could evaluate the claims, given that the Zika-microcephaly connection endorsers had a pressing sense that the public needed the information without delay and that their scientific discoveries would not be taken seriously otherwise."[180] Meg Stalcup, anthropology professor, Ottawa

The "little guys" (like Drs. Kleber Luz and Carlos Brito) checked a few sympathy boxes for journalists aligned with their politics.

"Doctors in the politically marginalized and poor northeast identified the circulating infection as Zika from its clinical presentation. While scientific competition is par for the course, and some northeastern experts did become widely recognized, ... clinicians on the ground collaboratively solved this mystery."[181] Meg Stalcup, (ibid)

Along that populist divide, the actual science gets somewhat left behind. There is essentially cheerleading or applause for the "little guys", the primary care doctors out in the field's winning out over the academic researchers in larger institutions. There are other dualities wherein the "winner" aligns with modern progressive identity causes (women v. men, poor v. rich, North v. South, downtrodden v. privileged, spiritualists v. academics, intuition v. the

scientific method, preemptive declarations v. deliberation, abortion v. the church).[182]

And that's just Brazil's own set of internal political dynamics. These may have supplied the tinder, but the match was lit externally, principally through Donald McNeil's New York Times' (self-admitted) driving the story directly to the most viewed web- or paper- pages, repeatedly.[183] This pattern began with their Rio reporter Simon Romero's December 2015 piece, *"Alarm Spreads in Brazil Over a Virus and a Surge in Malformed Infants"*[184] being similarly elevated to "front-page news" from the back pages.

The New York Times

Alarm Spreads in Brazil Over a Virus and a Surge in Malformed Infants

By Simon Romero
Dec. 30, 2015

SÃO PAULO, Brazil — A little-known virus spread by mosquitoes is causing one of the most alarming health crises to hit Brazil in decades, officials here warn: thousands of cases of brain damage, in which babies are born with unusually small heads.

⬆

⬆

"*Simon's story was up on the Times's website on December 30 and on the front page the next day. **From that point on, the Times was driving the story forward.** We wrote about it frequently, and the stories were **often on the front page** and prominently displayed on the website and mobile platform.*"

The Zika Epidemic, 2016, Donald McNeil

Within that first major article, December 2015,[185] there were two skeptical voices:

- "*Why this may have happened in Brazil and not elsewhere is at this stage difficult to answer,*" -- virologist Alain Kohl

- "*Zika virus doesn't worry us. It is a benign disease.*" -- Brazil's Minister of Health Dr. Arthur Chioro

But in short order, this minister, as seen (via this quote) as "dismissive", was himself dismissed. So much for skepticism. Early on, Brazil's health minister of communicable diseases **suggested women postpone having children (!).**

- *"If she can wait, then she should,"* said Mr. Claudio Maierovitch

This, despite saying a month later in front of the WHO in Geneva, that of the *"4,000 cases of microcephaly in the country, **only six** of the cases have been strongly linked to Zika virus via laboratory testing that confirms genetic material from the virus is present in the infant."*[186]

Brazil was in a near panic. President Rousseff saw "getting tough with the disease" as an opportunity for political vindication to stave off impeachment.

Within days, erstwhile **NYT** science reporter Donald McNeil[187] took up the torch. Professionally aware of his audience's sensibilities, Mr. McNeil tapped into readers' fears: aligning the mysterious disease, Zika, with their pre-existing "climate change" -concerns. He didn't focus on the problem in Brazil, rather on the threat to North Americans: bundling Zika with other *"scary bug-borne illnesses"* (Lyme, West Nile, Chagas, dengue, chikungunya)[188]; announcing its arrival in Puerto Rico; and claiming *"factors in the new spread are, for now, **unstoppable**"* (!).[189]

GLOBAL HEALTH

U.S. Becomes More Vulnerable to Tropical Diseases Like Zika

By Donald G. McNeil Jr.   Jan. 4, 2016

Tropical diseases — some of them never before seen in the United States — are marching northward as climate change lets mosquitoes and ticks expand their ranges.

Yikes, we're doomed!

Mr. McNeil offered high visibility to CDC-officials if they'd push for declaring an emergency travel advisory quickly. This seemed to do the trick.

> *"From (December 30) on, the Times was driving the story forward. We wrote about it frequently, and the stories were often on the front page (or) prominently displayed. I kept asking Tom Skinner why the CDC wasn't issuing a travel alert. Didn't the agency have a dengue-fighting operation in Puerto Rico? It was now on the front lines. Could I talk to the head of*

it?" Donald G. McNeil Jr., in his book, "Zika, The
Emerging Epidemic"

Mr. McNeil pushed the Zika-microcephaly narrative,
mightily, from his bully pulpit -- and then literally "wrote
the book" – whose sales benefited from the aggrandizement
of a Zika pandemic.

Of course viruses have no gender, class, race or politics. If
there is a toxic mechanism, *in utero*, bringing about
congenital deformity – that, too, has no political identity-
trait. It's not entirely impossible that our human
application of desired "winners" and "losers" in this (Zika
or microcephaly) clinical foot-race helped to cloud the
vision of the actual medical/diagnostic situation.
Furthermore, research institutions exist as institutions
because the "right" or "wrong" of a particular scientific
circumstance (should) have nothing to do with the
populism behind the investigators, inventors or vectors.

The New York Times and other media outlets' efforts were
successful. *"Zika"* -searches exceeded those made for
"Hillary Clinton"– even at her crucial juncture in the Iowa
caucus portion of the presidential election cycle.

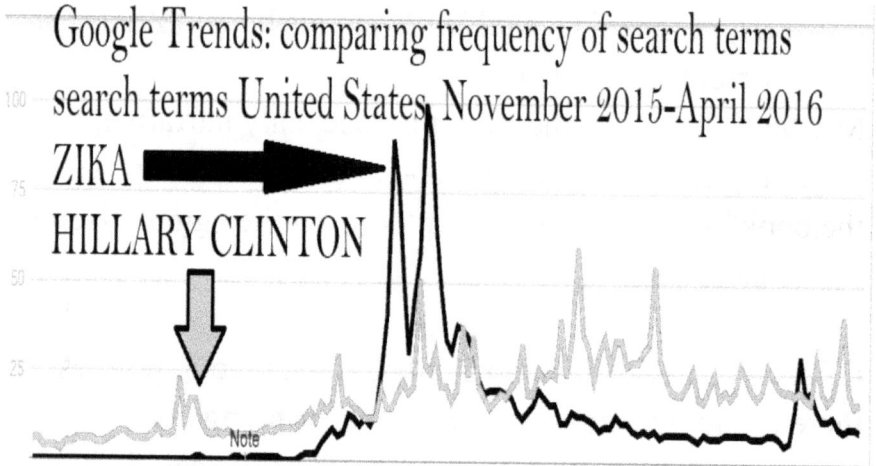

Google Trends: comparing frequency of search terms
search terms United States, November 2015-April 2016

ZIKA

HILLARY CLINTON

In the United States, this disease remained a full continent away at its apex of interest. Whether that interest was warranted is another story– but the New York Times' effective campaign of creating a storm of Zika-microcephaly fear did translate into a $1 billion expenditure by U.S. Congress[190], nine months later. The funding was delayed by congressional debate over (you guessed it) abortion. The issue of contention was whether a Zika expenditure would include government-paid abortions through Planned Parenthood.

Chicago Tribune
Congress lets abortion politics derail Zika fight

Commentary, July 5, 2016

By Arthur Caplan and Kelly McBride Folkers

Theories of Zika's Disappearance Abound

The PRESENCE of an ABSENCE

Humans (understandably) concentrate and focus during emergencies. Their actions are forced, frenetic, and perhaps somewhat ill-conceived, but accepted because of the potential downside of doing too little. We rightly err on the side of caution. But AFTER performing the quick expedient, do we ever do the less exciting work of looking back and assessing whether our cautions were necessary?

Earlier in the book we mentioned Dr. Seuss' Zax creatures (who will stubbornly only go forward). Cruise ships and trains have similar issues: the linear path is the most expedient; U-turns can be either cumbersome or impossible. A research-investigator's initiating a pivot requires institutional freedom and a mindset in short supply in academia . Research universities' oxygen is continued funding– which is predicated on surety of the previous set of results and novelty for the next. Consensus ruffles far fewer feathers than challenging the prevailing narrative, especially one which has contributed to your colleagues' salaries. Threaten this and you become a pariah on campus or a fired professor, off.

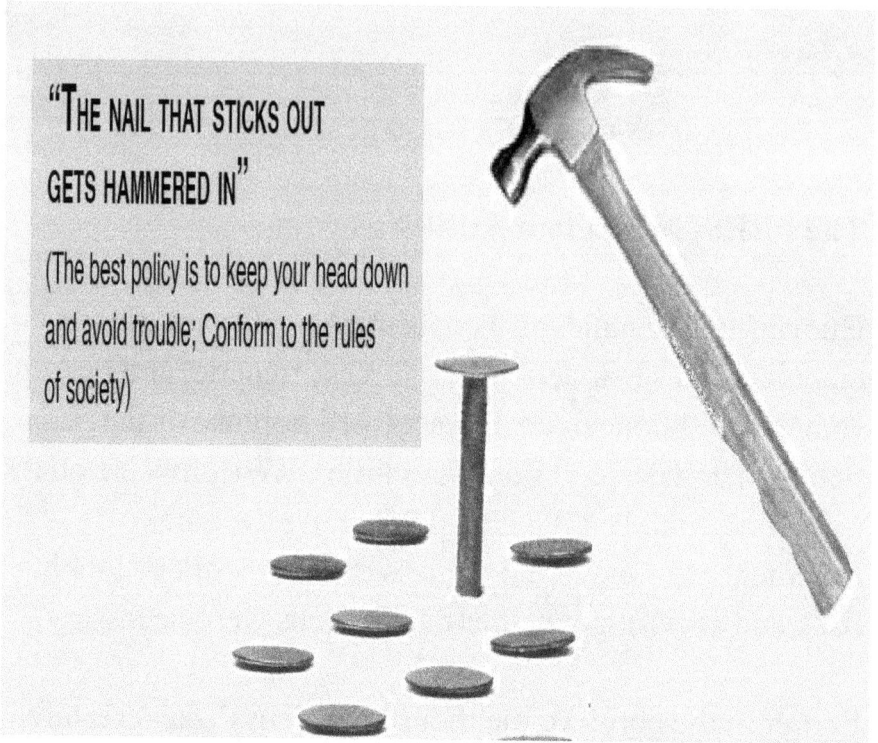

"**THE NAIL THAT STICKS OUT GETS HAMMERED IN**"

(The best policy is to keep your head down and avoid trouble; Conform to the rules of society)

An investigator's insight and acumen are tied to seeing and considering what others cannot. Fictional crime investigator Sherlock Holmes famously and endearingly solved one mystery essentially by spotting the **PRESENCE** of an **ABSENCE**, noticing (where no one else had) "*the dog that didn't bark in the night*"[191]:

> Gregory (Scotland Yard detective): Is there any other point to which you would wish to draw my attention?
>
> Holmes: To the curious incident of the dog in the night-time.
>
> Gregory: The dog did **nothing** in the night-time.
>
> Holmes: **_That_** was the curious incident (!)*

*(the dog was quiet from his familiarity with the thief)

Science writer, Maria Konnikova, advises : *"pay attention to what isn't there, not just what is."*

> *"(Holmes' dog) actively chose not to bark. The result of the two lines of reasoning is identical: a silent dog. But the implications are diametrically opposed: passively doing nothing, or actively doing something. Non-choices are choices, too.*
> *Take the well-known* **"default effect"**: *more often than not,* **we stick to default options and don't expend the energy to change**, *even if another option is in fact better for us.* **It's simply easier to do nothing.** *But that doesn't mean we've actually not done anything. We have.* **We've chosen, in a way, to remain silent**.*"[192]*

Can we deduce and derive something similar in the aftermath of the Zika epidemic of 2016? Have Zika-researchers, reviewers and journalists succumbed to the "default effect"? From the vantage of a much quieter 2019 (a few years back) with Zika absent and no microcephaly-pandemic (or Covid), the ghost of Sherlock Holmes had me wondering,

- *Had Zika really been all that it was claimed to have been?*
- *Where had it gone?*

- *What of the unrealized predictions of millions of microcephalic babies?*
- *Why is nobody clamoring for a re-examination, retraction, or reformulation?"*

These initial questions were followed by others:

- *How had the theories come about, and from whose hypotheses?*
- *Had they been validated by research and scientific institutions prior to being publicized?"*

The complete failure (*thankfully!!*) of any of the theories' dire predictions to come to pass must surely force scientists to go back to reconsider the events and come up with alternate possibilities, including potentially that the original theory had been wrong in certain ways. Right? Well not really.

The Current State of Zika Research

Despite billions of research dollars spent and thousands of investigative papers created within the instantly busy Zika research field, all of the original concepts of the Zika–microcephaly connection seem fairly intact and unchallenged within the scientific community.[193] [194] [195] [196]

Overturning Zika culled the last 2.5 years' "Top 40" Zika-microcephaly studies (sorted by relevance, per Google

Scholar; link <u>here</u>). Research-minds are thorough and meticulous, and the topics and methods vary. Zika-microcephaly seems still to be a relatively thriving research endeavor. Here are some general categories (2019-present):

- search of Zika's mechanism or active agent causing birth defects;
- attempting to find other congenital syndromes potentially caused by Zika;
- curious about any additional cofactors, other illnesses that might have brought the greater concentration of microcephaly in Brazil's northeast (2015) nearly exclusively
- trying to duplicate Zika's presumed effect, but in mice or other animals
- Looking at other abnormalities that might coincide with microcephaly
- Looking at later neurodevelopment in the affected cohort of microcephalic children
- Cautioning against any relaxation in the hunt for the "next Zika-microcephaly" epidemic, especially in the absence of any having existed in the meanwhile.
- Reviews of previous data, to tease out correlations.

Here again we can notice the "presence of an absence", there are no challenges to the underlying Zika-microcephaly theory, multiple years into the complete disappearance of microcephaly increases referable to Zika, worldwide, and in particular at the "scene of the crime" Brazil.

OBJECTION:
"Your Honor, the question assumes facts not in evidence."

The most intriguing of all recent Zika-microcephaly studies is Dr. Oliver Brady's retrospective analysis of over 4 million births within Brazil 2015–2017.[197] This ambitious study was organized to come to terms with the disappearance of Zika-microcephaly post-"pandemic" ("the presence of an absence" as *Overturning Zika* might put it); however, the entire project is in a sense doomed from the start from accepting as fact the unproven 2015 premises:

- "above average rates of the congenital malformation microcephaly in Northeast Brazil."
- "the first introduction of Zika in the Americas", and
- "evidence of the association between Zika virus infection in pregnancy and microcephaly."

The researchers point out

> "considerably lower rates of microcephaly were later observed elsewhere in Brazil and in other areas to which Zika subsequently spread. This led to suggestions of alternative causes or cofactors present in Northeast Brazil and uncertainty over the magnitude of risk that women infected with Zika in pregnancy faced."

From the vantage of this extensive look-back at 4 million births' data, they didn't find any "alternative non-Zika causes of the microcephaly outbreak" including dengue infection or previous yellow fever vaccination. What they did find was confirmation of the original assumptions, that there was more "Zika microcephaly"; but, only in the specific areas that had without proof claimed Zika-microcephaly, and only for the duration when they claimed it was there. If this sounds circular, that's because in fact it is. Partially the researchers acknowledge this:

> ***"Our analysis was limited by missing data prior to the establishment of nationwide Zika surveillance***, *and its findings may be affected by unmeasured confounding causes of microcephaly.... The finding of no alternative causes for geographic differences in microcephaly rate leads us to hypothesize that the **Northeast region was disproportionately affected by this Zika outbreak."***

Dear reader, what is an alternative explanation? Dr. Brady is saying that there was no prior data from which to make a comparison for 2015's phenomenon. He cannot explain why other areas with Zika had much lower occurrences of microcephaly that very same year. His conclusion is not that the Northeast (very much understandably, amidst panic) exaggerated its own claims – but rather that there was some vulnerability physically specifically for those Northeast citizens infected.

Credit should be given for the comprehensive nature of Dr. Brady's study. For this next diagram, each differently-shaped box documents the dataflow and analysis.

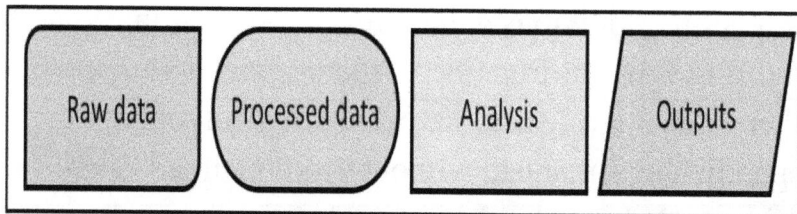

At the center of the diagram are the total birth numbers sifted through to discern the cases of microcephaly with structural brain defects (MWSD) in each category.[198] There are different levels of exposures, and naturally there is always some baseline microcephaly. Notably out of **millions of pregnancies** with varying levels of exposure, there are **only at most a few thousand cases of microcephaly** over the course of 2015, 2016, 2017; precisely the years Brazil's population was presumed maximally to have been exposed to Zika.

RAW DATA INPUT

1. (live births)
2. (microcephaly)
3. (PCR diagnosed dengue, chikungunya and Zika)
4. (confounders and bovine exposure)
5. (contaminated drinking water exposure)

Summary exposure

6,860,743 births
(2,791 microcephaly)

Full exposure

5,360,851 births
(2,282) microcephaly

3,603,823 births
(530 microcephaly)

ANALYSIS

1. Candidate cause models
2. Other birth defect models
3. Time-specific exposure models
4. Mapping microcephaly

Outputs

1. Most likely cause of microcephaly outbreak
2. Identification of other congenital impacts
3. Risk disaggregated by stage of pregnancy
4. Estimates of relative and absolute risk
5. Estimates of true Zika case burden

from: "Zika virus infection and microcephaly in Brazil 2015–2017: An observational analysis of over 4 million births" *PLoS medicine 5 Mar. 2019,*

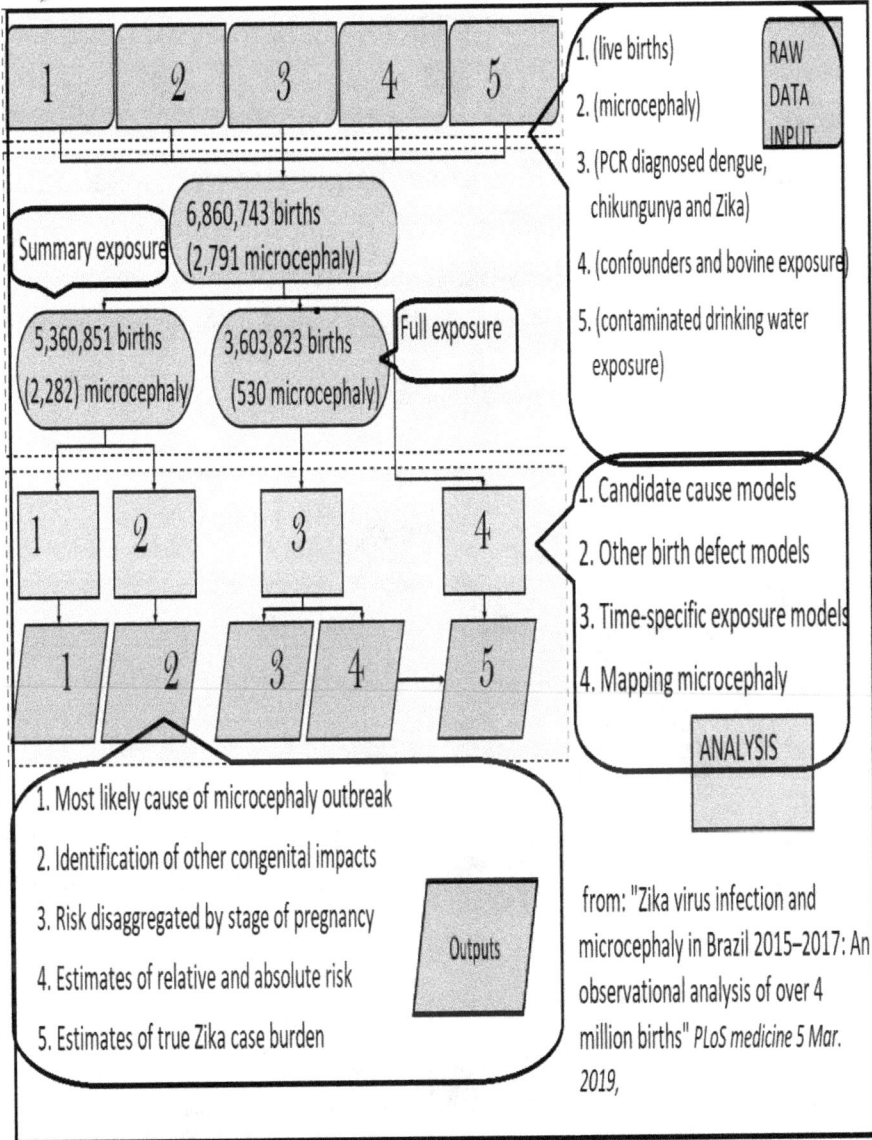

Dr. Brady's conclusion basically is that (if indeed Zika brought microcephaly, at all) it did so "ONLY" in Brazil's Northeast – although it is not so plainly stated. Somehow the geographic proximity to Recife brought microcephaly in greater numbers than elsewhere where there was similar exposure to the Zika-virus and Aedes aegypti.

Dr. Brady's map provides great insight into perhaps the real "epidemic". The highest risk of microcephaly occurred at urban areas in the Northeast, slightly related to but not fully coincident with the overall Zika incidents or claimed Zika-cases.[199]

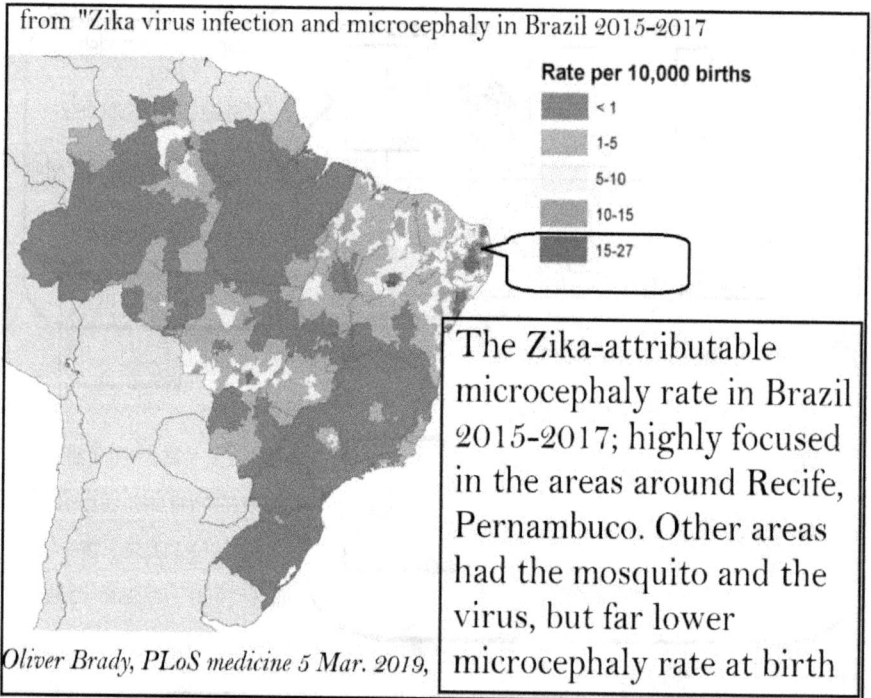

from "Zika virus infection and microcephaly in Brazil 2015-2017

Rate per 10,000 births
< 1
1-5
5-10
10-15
15-27

The Zika-attributable microcephaly rate in Brazil 2015-2017; highly focused in the areas around Recife, Pernambuco. Other areas had the mosquito and the virus, but far lower microcephaly rate at birth

Oliver Brady, PLoS medicine 5 Mar. 2019,

Overturning Zika agrees that "Northeast region was disproportionately affected"– but essentially and only as echo-reverberation of an unsubstantiated set of original claims: 1. Zika's existence, 2. Microcephaly's increase,

then 3. Zika as microcephaly causation. Panic brought hysteria, hysteria brought an enormously elevated number of claims– all against the backdrop of multiple different metrics of measuring head circumference and ZERO pre-existing federal or local registry of microcephaly data.

Overturning Zika would (respectfully) ask the authors to review the information within this book and then perhaps re-ask the same questions. Additionally we would ask about the flowchart's box: **"Full exposure history"** – how cases from 2015 could be included in this estimate without ever having been PCR tested in real time for Zika.

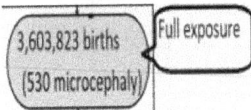

3,603,823 births (530 microcephaly) / Full exposure

"Yeah! That's the Ticket!"

Not just Dr. Oliver Brady's, but every prior research author's questioning why Zika did not produce microcephaly elsewhere (or later within Brazil) has been answered only by reinforcing (or "doubling-down on"; never doubting) the original premises; reminiscent of the children's story of the lady who swallowed a fly – having to make successfully larger errors in attempting to cover over the original.

Invariably, in Zika's case, this results in needing to apply special (and unprovable) retroactive rationales. Here are a few:

- In 2020, Dr. Louis Lambrechts at the Institut Pasteur noticed "*Zika's never caused a major human outbreak in (continental) Africa.*"[200] He began looking at the vector, mosquito Aedes aegypti, finding that "*Not all* Aedes aegypti *are the same*" and that Latin American and Asian types are more avid for biting humans harboring Zika.

  a) Even though Brazil's Aedes aegypti came from Africa in the slave trade

  b) Only Brazil had microcephaly. Not its neighbor Colombia, or other parts of Latin America with presumably the same mosquito

    c) 2018 Rajasthan India had Zika without microcephaly.

- In 2017, top echelon Brazilian **MoH-** and **WHO-** authors suggested Brazil's 2015 "first wave" of Zika may have been its **only** wave of Zika; proposing that the subsequent year Zika was mistaken for chikungunya. Basically, their theory is that **Zika was there in 2015 when it was never measured**, and in the subsequent year **when Zika was measured and announced as present, it wasn't**. This theory actually doesn't simplify things, and brings up in their view, three other possibilities:

    a) *"In 2016, there was herd immunity against ZIKV infection... (and) chikungunya infections were evidently misclassified as ZIKV infection (in Pernambuco in 2016 (personal communication from Dr. Carlos Brito(!)). "*

    b) *"**ZIKV infection during pregnancy doesn't bring microcephaly unless there is the presence of some other unknown cofactor** (somehow present in 2015, but absent in all following years)."*

    c) *"Fear of the adverse consequences of ZIKV infection led to fewer conceptions or a greater number of pregnancy-*

terminations in 2016. However, since any changes in the number of live births would be small, this hypothesis cannot be the principal reason why few cases of microcephaly were reported in the northeast region in 2016."

- In 2017, Yale's Dr. Albert Ko imagined that the **MoH** and **WHO** authors' mysterious "cofactor" was *"prior exposure to dengue (enhancing) the risk of birth defects from Zika."*[201]

  a) NB (<u>as covered in the previous chapter</u>), this doesn't apply: dengue frequency didn't vary between normals' and microcephalics' mothers.[202]

  b) Ko's research in the interim has not documented proof for this theory.

- In 2017, researchers in China suggested *"a mutant (Zika) strain particular to Brazil"* caused the microcephaly birth defects.[203] Dr. Anthony Fauci chimed in: *"This is a very good study and it gives a plausible explanation that is scientifically based"*; however,

  a) This theory has not borne fruit, given that this particular Brazilian strain never did the same damage in Brazil ever again. That would imply that the entire country had instant, total Zika immunity to this particular strain after only one year

b) The various Zika strains show minimal physical -- and no significant functional -- differences one from the other.

"direct comparison of sequence between the older and newer strains show only minimal <1% sequence change in ZIKV genome, none of which have been demonstrated to alter function; no conclusive evidence for differences between different ZIKV strains, regarding neural tropism or cell death"[204]

c) In 2019, Yale's Dr. Albert Ko (et al.) took the opposite approach (after the absence of microcephaly from India's Zika wave). *"Until new evidence suggests otherwise, all Zika strains should be considered to have the potential to cause birth defects, and many region-specific factors need to be taken into account when evaluating the population risks."*

- There is at least one published theory (that accepts the spike in microcephaly in Brazil, but...) does not attribute the damage to Zika – rather *"the alternative possibility that the use of the insecticide pyriproxyfen for control of mosquito populations in Brazilian drinking water is the primary cause."*[205]

To summarize, here are some of the retrospective "fixes" (or "guesses") attempting to explain away the inconvenient fact that 2015 Zika-microcephaly was in all aspects an unconfirmed event's occurring one time in one region and never reproduced anywhere else, including Brazil. These all seem improbable:

- Different versions of Aedes aegypti vary in Zika-delivery efficiency.
- When Zika could be measured and everyone was told Brazil had Zika 2016, it wasn't. It was chikungunya.
- Zika was still there in 2016, but nobody got it or microcephaly then because the entire population was immune after the single 2015 experience.
- No microcephaly increase was seen in subsequent years because people had stopped conceiving or continuing pregnancies (blatantly untrue although there had been some mild reduction).
- Zika needs some mysterious "cofactor" to perpetrate microcephaly. That cofactor was there when authorities weren't measuring microcephaly accurately nor Zika at all in 2015; yet, in the following year, when people had come around to the ability to do both, the cofactor had disappeared. Ditto, for every year thereafter, no Zika, no microcephaly.
- Previous dengue exposure is that "cofactor" (data do not back this up).

"a mutant (Zika) strain particular to Brazil" caused the microcephaly birth defects.

From my vantage, these seem like stretching reality to fit an agenda; explanations more and more grandiose to paper over the underlying weakness of the original proposition. *Nobody, nobody*, and I mean *nobody*[206] seems able to voice doubting any or all of the four Zika-conjectures, all of which need to hold true for the Zika-microcephaly connection to be real.

1. That certain, completely-untested cases essentially synonymous with dengue, showing up in dengue-endemic areas were instead and definitely a never-before-seen-in-Brazil Zika.
2. That this Zika, previously eternally harmless to humans, had a heretofore-unobserved dark side
3. That claiming more microcephaly (without prior data-comparisons) in one area of Brazil (during a panic) implied a nationwide increase,
4. And that one improbability had caused the other.

Reputations are tied up in theories, and as with Zika research, there can be a certain forward-momentum in carrying on resolutely with a theory even if it is not fulfilling all of its obligations in doing what a theory should do: predict (reproducibly) future events. There's no reason we shouldn't expect scientific theories to bear out in real life.

If someone should postulate a "theory of gravitation" wherein heavy objects, untethered, float upwards, a simple experiment of dropping a stone or coffee mug to the ground is more than sufficient to refute it. Likewise, many of the predicted events pertaining to the Zika virus simply never occurred. Yet somehow the theory still holds, in some ways now stronger and more entrenched than before.

Clearly epidemiologists have larger and more contemporaneously pressing events on hand and on their minds (Covid 19), so perhaps it's too much to ask for a retrospective. **Overturning Zika** hopes to fill in that gap.

The Experts Respond

It's July, 2019 and the latest greatest pandemic is only just emerging in Wuhan, China. With this sense of "dramatic irony", let's look back and see how our previous great pandemic, Zika, then appeared to experts: basically, as something from which not to let go:

The New York Times

The Zika Virus Is Still a Threat. Here's What the Experts Know.

By Andrew Jacobs
July 2, 2019

... seemingly overnight, **the epidemic evaporated** and public attention moved on. **But Zika, it turns out, did not vanish.**

"Zika has completely fallen off the radar, but the lack of media attention doesn't mean it's disappeared," said Dr. Karin Nielson, a pediatric infectious disease specialist at U.C.L.A. who studies Zika's impact in Brazil. **"In some ways, the situation is a bit more dangerous because people aren't aware of it."**

"The next outbreak is not a matter of if, but when," said Dr. Ernesto T.A. Marques

Dr. Eve Lackritz, who leads **W.H.O.'s Zika Task Force, said one of her main tasks is to keep up the sense of urgency**. "My biggest fear is complacency and lack of interest by the global community," she said.

Don't let the previous three years of utter absence of Zika - related microcephaly or Zika in general as an illness to be feared fool you, citizen. This is only the penultimate scene of the horror movie. You only think you've killed off the monster but he's got you right where he wants you before reappearing for the final gruesome battle. The WHO's Zika Task Force physician leader's admitted main task is "**to keep up the sense of urgency.**"

- The CDC's Lyle Peterson worried *"the world is unprepared for the next outbreak."*[207]
- Brazil's Paolo Zanotto's *"biggest fear is that we will never get rid of Zika."* [208]

This fear-mongering "logic" is self-perpetuating; moreover, it's counterproductive in the actuality of "preparing for the next outbreak": diverting resources to fighting the last battle – with the next one's being completely different. Instead of trying to "**keep up the sense of urgency**" for nonexistent Zika, more agile, forward-looking thought-processes might very well have prevented Covid's leak – over which the CDC bears complicity, having furtively supported Wuhan's Institute of Virology's "gain of function" SARS (coronavirus/Covid) research.

The Next Zika pandemic? Coming Soon!?

In personal communication with one top academic Zika-expert (MD, PhD) with professional roles both in Brazil and the US, *Overturning Zika* asked a series of questions,

is grateful for these thorough responses. A summary is featured below, (should you care to skip this section).

Overturning Zika asks:

> *"It's now 2021 and Zika (or perhaps dengue, cross-reactively) might be permeating the world -- but microcephaly, not so much. One million such births per year were estimated for the tropics' 3.3 billion people back in 2016, but overall microcephaly numbers don't seem to have increased once everyone arrived on the same page regarding standards of measurement. We've been told by the CDC and the WHO to remain on alert for the next outbreak, when will that be?*

The academic Zika-expert's response:

> *"Zika cases continue to appear in several regions in endemic levels without sparking an epidemic. Zika related congenital abnormalities continue to appear in small numbers.*
>
> *Microcephaly is only one symptom that is present in most severe cases but is not the most frequent. Retinal and cochlear defects are much more frequent but are only detected later in life. Follow up studies have shown that many children born from Zika-exposed mothers that were normal at birth are showing several types of complications.*
>
> *Diseases caused by flaviviruses, like Zika, have cycles. It is very well known that different dengue serotypes alternate outbreaks over time. Zika*

entered in this group and will enter in alternating cycles together with dengue serotypes. The great dengue epidemics in Brazil have happened in about 7-8 years intervals but different cycles occur in other regions. Southeast Asia has shorter cycles. The length of the cycles depends on environmental factors, effectiveness of the vector control programs and herd immunity.

In northeast Brazil, it is expected that a big dengue outbreak should occur in the next two years. Most likely Dengue 1 or dengue 2 since there is a greatest fraction of the population that is susceptible to these serotypes. Due to cross-reactivity between dengue, Zika enters this dynamic as if it were one more dengue serotype. So, imagine that one epidemic can appear every 7–8-year intervals -- and one of 4 dengue serotypes plus Zika will emerge based on the herd immunity. **[Our academic Zika-expert] estimates that we have about a 35% chance to have another large Zika outbreak in this decade, and this should grow with time.**

Overturning Zika followed up with this set of questions:

Doesn't dengue come in year-by-year waves?

Our academic Zika-expert:

Yes, every summer an increase of cases of dengue and other arboviruses evolve with an average cycle of 7 to 8 years. Zika was introduced in this context and circulated predominantly in regions whose previous arboviral outbreak was 2002's Dengue 3

outbreak. Other countries have shorter cycles of every two or three years.

Overturning Zika

"Are you saying that the difference in "variant" strains brings a cycle that would be the same for Zika?"

Academic Zika-expert:

"Dengue serotypes 1,2,3, and 4 are different viruses not variants. Although each of the serotypes have variants, they do not immunologically escape the immune responses of their serotype. The variants have an impact on the clinical outcome of the patient, but not in regard to herd immunity."

Overturning Zika

"Wouldn't it then have borne its full effect in the rest of the Americas' tropics which were like Brazil the prior year, naïve to Zika prior- giving them similar microcephaly rate increases? That didn't occur."

Academic Zika-expert:

The transmission of arboviruses depends on several factors, like environment, vector control programs, mosquito species, social economic factors and other factors related to herd immunity. High dengue antibodies provide protection against Zika. The levels of Zika transmission in each country were not well documented due to several reasons. 1) it is very mild and many people do not realize they have acute Zika, 2) The diagnostic kits cross-react with dengue and 3) Nobody was looking for Zika until very long after the outbreak. For example, in Northeast Brazil the outbreak was in Jan-June 2015 and all cases were reported as dengue. (It was only after the) January 2016 microcephaly outbreak, that there was the ability to diagnose Zika. So basically only a few sites were capable of measuring the real rate of infection. It is not necessarily true that the Zika virus would have been able to reach epidemic levels in all Latin-America just because the population is naïve. There were naïve populations in many places like the US or EU but the environmental factors did not allow it. " [i.e., carrier mosquito Aedes aegypti doesn't live outside the tropics]

Overturning Zika:

"Would not there have been microcephaly numbers' increase in places similarly naïve to 2015 Brazil, other places throughout the tropics, and the subsequent years? That didn't seem to have occurred."

Academic Zika-expert:

"They occurred, (but different countries had different standards for "confirming") cases of Zika Congenital syndrome. Some countries only confirmed a case if they had a PCR positive Zika-test in the baby, however the mothers had been infected 6 months before and the mothers would not still have showed positive at delivery. The number of suspected

Zika-microcephaly cases in Colombia is huge, but they only confirmed 10%"

Overturning Zika:

"It also seems a little bit "moving the goalposts" to take the focus off microcephaly, and expand Zika's supposed neurotoxicity to problems that can't be measured at birth and are indistinguishable from rubella's or other infections' or toxins' – which themselves, even better known, can't be automatically diagnosed either."

Academic Zika-expert:

"This is not the case at all, these cases were also investigated for other TORCH diseases and there is no ambiguity. There are two very large longitudinal cohorts of pregnant women, one coordinated by the NIH and other by the European Union each with about 8000 participants. They started the follow up during the first trimester of gestation until the babies reached 1 year old. All cases of children exposed and an equivalent number of non-exposed to Zika will be followed up. Many parameters are being measured and there are studies of other TORCH diseases in these same cohorts."

Overturning Zika:

"Isn't current microcephaly (at least as a given measurable at birth) ready for comparisons with Brazil 2015's presumed epidemic and/or with Zika-positive maternal testing (which, of course, was not *actually measured in the 2015 Brazilian presumed epidemic)?"*

Academic Zika-expert:

*"The association is epidemiological and biological-- not only based on serology, but also with the identification of the virus on fetus brains and amniotic fluid. Animal models also show that the virus causes microcephaly and neurodevelopmental problems. There are cases in which it was compared to different strains, and the most pathogenic were the Brazil and Puerto Rico strains. By the way Puerto Rico also had a high number of cases. **However, there are some factors not explained. In Recife, the epicenter of the outbreak, 97% of the children with microcephaly were from low socioeconomic status**. However, the seroprevalence of Zika in these groups is (higher). Several environmental and nutritional cofactors have been explored but so far none of those have been associated."*

Summary of the academic Zika-expert's Q&A

Disclaimer: ***Overturning Zika*** is making this summary under the premises of its multiple prior chapters. This is ***Overturning Zika***'s summary, not the academic expert's. Within this summary, we aim to give the gist of what was said above along ***Overturning Zika***'s rebuttals. It is hoped that the curious reader will be able to discern channeling the expert's voice and ***Overturning Zika's*** criticisms.

- Zika may still be out and about in the world but not in any concentrated epidemic fashion anywhere. The definition of related congenital abnormality seems to be expanding to things that overlap with rubella-syndrome and/or other toxins or infections. Children are being looked at later in age, not just birth. Of course this creates new "blurred lines", harder-to-study populations from whom to attribute causes, years after.

- Our expert states that dengue has different subtypes and these seem to go in cycles in Brazil, much like the flu does elsewhere. Cycle lengths differ potentially in different regions. The causes aren't known but *"The Usual Suspects"* (environmental factors, mosquito control, herd immunity, socioeconomic status) are brought up when no immediate answers are available. So there is some suspicion that Zika is waiting for its "cycle" for the same "usual suspect" reasons (rather than perhaps it was never there in the first place and was mistakenly transposed, and in actuality a misdiagnosed dengue).

- The expert's view is that Zika will be one extra card in the deck (of four dengue subtypes) and that one of these cards becomes a big deal only every seven or eight years, so Zika might only show up as epidemic in 30 years, or as stated here a one in three chance within a decade. That's a pretty big cushion behind which to hide as far as having to deny the original

Zika-premise. Are we going to wait 30 years to refute this?

- The expert's view is that Zika-related microcephaly was not counted or shown to exist throughout the rest of Latin America's tropical area(s) because the other countries didn't want to promote or permit such numbers. The expert feels that Colombia's numbers were "huge" but remain hidden because of reluctance to "confirm" microcephaly (*Overturning Zika* infers from these answers).
- There will be some large studies that might help separate Zika from the other TORCH congenital birth defect causation infection agents. The expert's view is these studies may find other congenital Zika-issues, rather than simply microcephaly. *Overturning Zika* feels this is adding certain distracting features to the simple question: does Zika cause microcephaly? The answer lately has been "no", but now it's "look over there", maybe it does some of the things that rubella has long been known to do – and mind you, that Zika's near-twin dengue has NEVER done.
- Even though we (in academia) are not seeing clinical microcephaly, we are falling back to the position that some animal studies show passage of Zika across the placenta and into neurodevelopmental cells, including the brain.
- Additionally (although *Overturning Zika* notes that the "mutant strain"-theory has been elsewhere

discredited), our expert emphasizes Brazil and Puerto Rico had the "super strain" that caused more microcephaly. Presumably that explains why other places haven't gotten the microcephaly. It doesn't explain why Brazil didn't get it later itself.

- Despite mentioning "the usual suspects" (of microcephaly-associated causes, listed above), the only one of these that's really been shown to be associated with microcephaly is **poverty.**

Overturning Zika reiterates that numerous other microcephaly-causes are associated with poverty, to the extent that in the sense "poverty" could be listed as a cause, itself. Why that's not the microcephaly rationale instead of "Zika", *Overturning Zika* can't say. And of course all of this theorizing works under the premise that there was Zika, and that there was in fact an increase in microcephaly – both of which were completely unverified in the pandemic year.

A Shorter Expert Exchange

Overturning Zika had a separate discussion with an academic biostatistician (PhD) who had analyzed Zika – later admitting:

> *"The reality ended up closer to our lowest estimates in (societal disease) costs. There might have been **an excess of panic produced by media reports in (Brazil's) Northern states;** however, in global terms*

the impact of Zika has been understated -- because of difficulties in assigning health conditions to Zika, and because most of the cases happen in tropical countries where many diseases are neglected. (Follow-up confirmation is not anticipated, because) Covid-19 currently fully preoccupies epidemiologists.)"

Overturning Zika's quick summary of this interaction:

- even somebody mathematically oriented (a biostatistician) has trouble separating facts from beliefs.
- Zika completely underwhelmed, fell far short of the team's estimates for wreaking havoc, damage, chaos, and microcephaly outside Brazil – YET the small step of acknowledging 2015 Brazil's panic and exaggeration is quickly counterbalanced with the nebulous excuse that Zika is hard to ascertain because it's not a distinct entity and/or because it happens in places where people don't follow up with diseases.
- This is an argument that of course if true would make one doubt the original 2015 Zika-microcephaly premise, biostatistically and in reality, as well.

The Broken Promise of the Zika Vaccine; Women's Lives in Limbo

What Is the Opposite of *"Warp Speed"*?

The Covid-19 vaccine represented a genuine technological innovation: a brand-new formulation for a brand-new disease (Covid-19), accomplished quickly. This vaccine was meant to facilitate an exit from lockdowns and similar measures. Arguably, the politics behind the mass enforcement of vaccination has been heavy-handed, but there has been *"product"*. While **Operation Warp Speed**[209] achieved its goal within months– conversely a Zika vaccine remains still an unrealized ambition, these many years later.

Under the threat of Zika microcephaly (2015-2016), Brazil and the tropics worldwide lived in fear. In Colombia, El Salvador, Jamaica and parts of Brazil, government ministers (advised by WHO public health experts[210]) [211] told the populace to stop having children (awaiting a vaccine).[212] The decline in births implies that hundreds of thousands of women listened – at least for a time. Seven years and no Zika vaccine later, those women who dutifully obeyed are worse off.

The sudden and complete disappearance of both Zika-illness and Zika-microcephaly undoubtedly slowed Zika-vaccine's impetus. Vaccines entail some risk and our epidemiologic warriors in public health seem loath to start a vaccine-trial skirmish when nobody is suffering from the virus. Side effects with no countervailing benefits is bad vaccine-PR.

2020 was a year lived in fear. Covid-19 had been unleashed from China, with inadequate information of China's experience. Perhaps over-cautiously, we hunkered down in lockdowns premised upon an impending vaccine's validating that strategy.

Operation Warp Speed essentially and nearly miraculously lived up to its name, producing a vaccine in less than one calendar year. It was an unprecedented *tour de force*, combining the speed and innovation central to business – with the funding, bureaucratic reach, and liability-reduction of government.

The Covid-19 vaccine forecloses against the return of "ancestral" Wuhan Covid 19; while less effective against *Greek-letter* (α, β, Δ, o) variants[213] (which fortunately trend milder, sequentially). Ideally, a vaccine precedes a pandemic and is changed out for new strains, as with yearly-differing flu-shots.

In business, the formula for success classically has been to *"under-promise and over-deliver"*, setting reasonable expectations – and then knocking them out of the park. The

Covid-19 vaccine's over-promised lockdowns' and restrictions' disappearance-- yet ***under-delivered***, dissatisfying many.

With each new variant strain's increasing physical and genomic difference from the original, the vaccine becomes less of a match; nonetheless, the Covid-19 vaccine was repeatedly pushed for "Covid-20" (α, β) and "Covid-21" (Δ, o) variant-strains.[214] [215] *"Breakthrough"*-cases are mischaracterized in that Covid-19-vaccinated individuals are not much protected against omicron, in the first place. No one should be shocked about getting influenza only having been vaccinated for the strain from two years prior.

The Project Sputters

The Covid-19 response story (as with all human endeavor) is one of imperfection: overlong lockdowns and masking – but at least *its* vaccine exists. The Zika-response by this standard is a failure– especially given this severe and incredible demand made to women, *"The rapid spread of Zika ... has **prompted governments to do something without much precedent in human history: urge people to hold off on having kids.***[216]

The Atlantic

A Country Without Babies

Governments are telling women to postpone pregnancy during the Zika outbreak. What happens if they do?

The science was certainly adequate at the time to have created a Zika vaccine. Money was no issue; billions of dollars were committed to the fight against Zika. Other flavivirus illnesses (Japanese encephalitis and Yellow Fever) had previously been vaccine-formulated.

Sanofi's CSO Dr. Gary Nabel, formerly of NIH's Vaccine Research Center optimistically said (2016),

> *"What causes you to ... really launch a vaccine program is when you see that there are serious public health effects. The Zika virus qualifies (sic). Sanofi has a vaccine for dengue.* [217] *Dengvaxia (may) cross-protect against Zika virus (although) concerns related to pregnancy (make it) preferable to use inactivated rather than attenuated (live) virus as in Dengvaxia."* [218]

Work began on Dengvaxia in 2009, but its implementation was slowed by controversy.

Children in the Philippines (2017) suffered more severe disease following vaccination,[219] precisely the opposite of a vaccine's desired result. [220] Dengvaxia was nonetheless globally licensed and released in 2019.

Coincident with the Zika microcephaly panic, researchers were gung-ho, producing dozens of optimistic Zika-vaccine videos. Here are a featured three, including Dr. Fauci's.

Initially a flurry of Zika-vaccine videos, CBS News, etc.

2016 — DR. ANTHONY FAUCI NIAID DIRECTOR — Developing Zika Vaccine

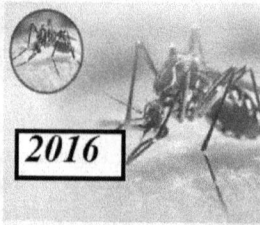

2016 — Human trials begin for Zika vaccine

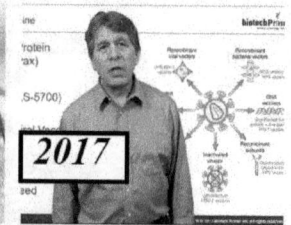

2017 — Making a Zika Vaccine

Only a single video produced since 2017 on this topic

After that first year, enthusiasm quickly tailed off. Think of earnest and excited New Year's resolutions, forgotten months later. That's almost impossible to replicate with a billion-dollar budget, reputations at stake (cf. Dr. Fauci's) and the world's full attention; but so it has come to pass.

Before the enthusiasm and emergency subsided, Dr. Fauci announced: *"We're (involved with) a human phase-one trial to determine safety and whether it induces a good response. It will likely do that starting sometime in the fourth quarter of 2016. So all of that is already in progress."*[221]

Considerations for Developing a Zika Virus Vaccine

Hilary D. Marston, M.D., M.P.H., Nicole Lurie, M.D., M.S.P.H., Luciana L. Borio, M.D., and **Anthony S. Fauci, M.D.**

(Article, plus exclusive) Interview with Dr. Anthony Fauci on potential strategies for conducting clinical trials of Zika virus vaccines.

The NEW ENGLAND JOURNAL of MEDICINE

September 29, 2016

Indeed the phase-2 trials began in 2017 in the United States, Brazil, Peru, Costa Rica, Panama and Mexico[222] amidst declining public interest in Zika (by 2017).

Dr. Fauci, despite stating Zika vaccine remained *"a very high priority"* for the NIAID[223] essentially failed on the follow-through; NIAID had not lined up vaccine manufacturers. Moreover the entire foundation of the Zika vaccine project relied on gullibility: taking at face value the original Zika-microcephaly pandemic claims, without adequate CDC-, NIAID- or NIH- investigatory confirmation. As a result they were blindsided by the near-immediate disappearance of Zika. An absence of Latin American cases required massive expansion of clinical trials, entailing greater cost and bureaucratic difficulty.

In 2018, ten clinical sites were still plugging away on multiple vaccine-formulations including Moderna's mRNA-. The contemporaneous review in **"Vaccine"**, 2018, never mentions *"delay"*, *"late"*, *"problem"* or *"failure"*.[224] Seemingly everything remained on track (as long as the funding persisted) -- although the populace, under the presumption of a genuine Zika-microcephaly problem -- had received no respite or relief. People had not been given a vaccine nor told the initial claims were exaggerated

Infect People to Save Them?

Humbled by the Tuskegee experiment's exposure,[225] US agencies considering infecting human subjects must pass an ethics-review. The prospect of a Zika-vaccine triggered a bioethics symposium December 2016 in which it was said:

> *"Researchers have long recognized the ethical stakes in infection challenge studies. Walter Reed's yellow fever experiments in Cuba in the early 1900s provided protections for subjects. Researchers who conducted challenge studies with syphilis and gonorrhea in Guatemala in the late 1940s ... withheld established effective treatment, did not take measures to protect third parties from the spread of the infection, and inappropriately included subjects from vulnerable populations."[226]*

Tasked with the question of whether a Zika vaccine trial should purposely infect individuals in a "Controlled Human Infection Model" (*CHIM*) -study), the bioethics panel answered, "no!".

NIAID's & Walter Reed's (WRAIR) convened panel, questioning infecting healthy subjects with Zika

ETHICAL CONSIDERATIONS FOR ZIKA VIRUS HUMAN CHALLENGE TRIALS

REPORT & RECOMMENDATIONS

February 2017

Can a Zika virus human challenge trial be ethically justified? If so, under what conditions? *The answer was "NO!"*

"the committee has determined that (current) conditions (high potential risks, other pathways available) preclude the conduct of a Zika virus human challenge trial"

Lo and behold– with the clock ticking two years into US taxpayers' ~$1 billion in unproductive Zika research – the NIAID and bioethicists seemed to have a change of heart: **to go ahead and purposefully infect test subjects**. This might make some sense in a raging pandemic; however Zika-microcephaly *(inconveniently for researchers, fortunately for people)* remained absent as a medical concern anywhere on earth.

In the absence of an emergency, how does one explain this sudden change of heart? Desperation? Flop sweat? Fear of failure? The ethics hadn't changed, but the politics had. This 2018 Science-article's subheading gives a hint of the predicament:

As massive Zika vaccine trial struggles, researchers revive plan to intentionally infect humans

Science (September 2018)

Disappearance of Zika in Americas makes it tough for $110 million trial to evaluate worth of vaccine candidate

"A $110 million vaccine trial faces an unexpected, and ironic, challenge. Cases of Zika have plummeted to levels so low that most people vaccinated in the trial likely will never be exposed to the virus, which could make it impossible to tell whether the vaccine works."

"Right now, there are no infections, and certainly not enough to even think about an efficacy signal at this point," added Anthony Fauci[227]

The NIAID could have "declared victory" and gone home (essentially)– or at least put the studies on hold. Cataloged scientific efforts don't disappear, even if Zika does. If Zika, somehow, returns as a problem, those efforts may be revived with even more advanced future technology.

Instead, NIAID pressed ahead, erring on the side of potentially violating documented human-challenge ethical standards. Why is that? The answer might be found in comparing these two statements from NIAID's Brazilian

collaborator Dr. Esper Kallás of the University of São Paulo:

"It's a good dilemma because we don't have Zika anymore."

"But it's a dilemma. Everybody is concerned about it. It's a lot of investment."[228]

In business, this is known as the ***"Sunk Cost Fallacy"***
"The more our past investment, the harder it is to abandon it. We ignore the promise of a better experience in the future by making an attempt to negate a loss in the past. In other words, our past investments over-influence our current decisions."[229]

Why do we fall for sunk cost bias?

Admitting any mistake undercuts reputation

past successes give false hope based on information bias

Avoiding pain over gain

emotion over rationality

"quitting's for losers"

"There's this sense the epidemic will hit our region, but we don't know when," Kallás says. *"We don't understand why it didn't happen already."* The study's leader Dr. Anna Durbin was frustrated by the constraints of the bioethics committee, feeling that avoiding a Controlled Human Infection Model *"was a great setback. If we had been allowed to go forward, we'd know today which vaccine candidates look good."*

This very same *Science*-article shows "Zika's vanishing act".

Zika's vanishing act

Weekly counts of new Zika cases, suspected and confirmed, have plummeted in North and South American countries hosting a vaccine trial.

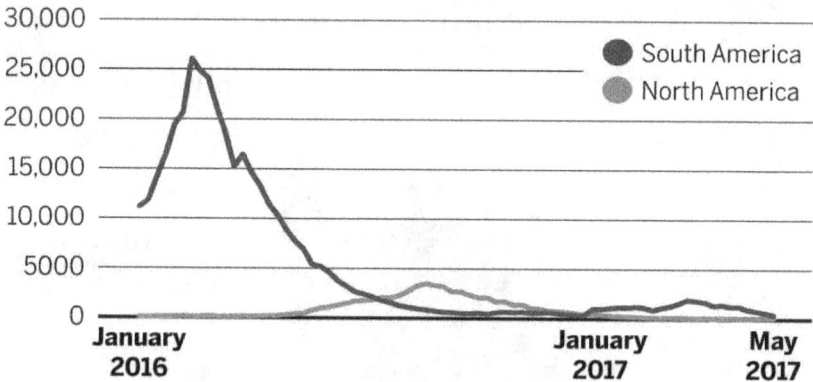

Mind you this was at the height of scrutiny. Numbers since then have been effectively zero continuously.

Zika's vanishing act

Weekly counts of new Zika cases, suspected and confirmed, have plummeted in North and South American countries hosting a vaccine trial.

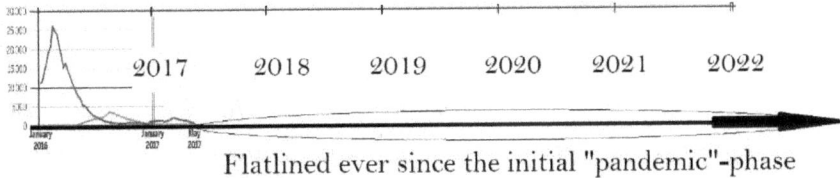

Flatlined ever since the initial "pandemic"-phase

If anything, there would be far fewer compelling reasons to go ahead with a CHIM, but the bioethicists read these same tea leaves quite differently.

Even after one more year of continued Zika-disappearance, the bioethics committee chair Dr. Seema Shah turned the 2017 ethics report on its head, saying *"There's a compelling reason to conduct a human challenge trial now (2018). The details are complicated and it's important to have a rigorous review."*

It is disconcerting to hear Dr. Shah say "the details are complicated". That would seem to be professor-speak for *"I can't really express what those reasons might be"* or *"whatever those reasons are, they are probably not strictly speaking ethical"*.

At the same juncture, Dr. Fauci added *"If they're careful, we have no problems supporting (a CHIM). Zika was a very ominous threat just a couple of years ago, and there is certainly the possibility that it is going to come back. It's a risk that you'll spend this money and never use the vaccine, but balancing the importance of this infection and the impact it could have, we felt it was a good decision to move ahead. And I would be happy to defend that anywhere."*

Johns Hopkins' Dr. Anna Durbin hoped *"to start injecting Zika virus into people immunized with a vaccine in 2019."* Brazil's "FDA", **ANVISA, negated** these fervent hopes of Drs. Kallas, Fauci, Shah, Durbin *et al.* **by forbidding a CHIM on Brazilian soil** (at a time of no Zika-microcephaly). Dr. Durbin remains undeterred and is organizing a CHIM on US soil, 2022 – six years (and counting) into Zika-microcephaly's continuing non-presence. No pandemic, no problem! The US-funded research continues!

And it continues at an unseen expense: shelving more valuable studies, research and projects merely because Zika-microcephaly had an outsized and undeserved presence in 2016's news-cycle. Dr. Fauci lobbied the U.S. Congress for $1 billion in Zika funding (2016) admitting:

> *"First, we took money from other infections (malaria, TB and Ebola); ... then from cancer, diabetes, heart disease and mental health ... to prepare for Zika vaccine trials."* [230]

Inasmuch as schools leverage cities by cutting the arts during money shortages, this budgetary jiu-jitsu succeeded in prying open Congress' purse– but at what ultimate cost?

This itchy trigger-finger approach to vaccination is not necessarily the wisest. If **Overturning Zika** is correct, Zika-microcephaly will never return, vaccine or not. If Zika-microcephaly is a genuine phenomenon -- upon its recurrence, technology will have improved, and all the existing science remains. The overall "good news" from Covid-19 is that mRNA vaccines can be made fairly quickly once the proper genetic coding has been sequenced.

Coding Coda

Moderna CEO Stephane Bancel noted (2021):

> *"The third chapter for us is public health and we are very happy that we are working in public-private partnership (for example, the US government) on getting a Zika vaccine to market. Zika has not left the planet. Zika is going to come back* **(sic)**. *We need to be ready when it comes back to minimize the potential human toll of this awful virus carried by mosquitoes."* [231]

Overturning Zika disagrees with Dr. Bancel's forgivably misconstruing Zika (i.e. everyone in the scientific establishment agrees with his viewpoint, currently), but his sentiment is reasonable and the company's technology will be essential for viral threats as they emerge. One hopes such future responses will be quicker in turnaround time and not duplicate the debacle of the missing Zika-vaccine.

Public health officials need to take greater care in their dire warnings which seem to err continually on the side of panic amongst the populace. Those unfortunate women who were told not to have children are left with the burden of absence. It is our fervent hope that they have managed to find happiness to replace that which was forsaken on this altar of (pseudo) science.

The Verdict

It appears that so many of Brazil's clinicians on the Zika trail were forthright, intrepid, brave, interested and curious—all excellent qualities for physicians. Many were engaged with seeing and treating patients, and had the proper empathy and concern for the distraught mothers and damaged babies. What they didn't have, and what they couldn't have had (and possibly didn't pursue), was access to a large-enough data-set to make a firm diagnostic-connection.

Partially this is because certain aspects didn't exist fully in Brazil at the time,

- Zika testing was not generally available and had not been performed clinically.
- There was no microcephaly federal registry; so, no baseline comparison data set.
- There were no firm uniform size-metrics for determining microcephaly at birth.

– but part of the blame is theirs as well. Scientific protocol was not followed. *"Science by press release"* did not allow ideas to percolate through the research academies' "checks and balances" process of conjecture, debate, proof, review. Leaks to the media may be framed as helping the general public– but arguably the premature release of the Zika-microcephaly theory more benefit to the researchers themselves than to the resultantly frenzied populace.

There was no real need to jump the gun in this way. If the theory was correct, current births' damage had been done seven months prior. No immediate maternal changes in behavior or mosquito exposure was going to help after the fact. Certainly it would be prudent to have newly-pregnant moms avoid mosquito bites, but such measures and advice are timeless – and could have been promoted without engendering panic.

At the junctures in which access to data and scientific review was available, there was no change to the theories by the proponents, or coverage much by the press. It seems almost as if this type of "fact checking" was disdained and avoided. People stayed very much "on message" in promoting the Zika-microcephaly connection, as shaky as it was.

Here is a telling paragraph from *Zika: From the Brazilian Backlands to Global Threat*. Dr. Brito may not have liked the epidemiologic surveillance figures, but to think that they "muddled the epidemics" is putting the cart before the horse – as if his theories are more important than data that might refute them.

> "from the time that Zika was found to be circulating in Brazil until October 10, 2015, Pernambuco one of the epicenters of Zika in Brazil — tallied **only four cases of**

ZIKV infection, as confirmed by the Evandro Chagas Institute. As Dr Brito saw it, the epidemiological surveillance figures effectively muddled the three epidemics (Zika, dengue, chikungunya?). (Dr. Brito) insisted on explaining something basic about clinical practice: *"Once an epidemic has been confirmed, diagnosis is clinical. The laboratory is only used in critical cases. (!?)"*
… Because the numbers were such a mess, Dr Brito was even more surprised when Dr Vanessa Van der Linden, the neuropediatrician from Recife (Pernambuco), called him on the afternoon of October 19, 2015 to talk about children with microcephaly."

At the time of the connection of the two theories, presence-of-Zika-as-illness and microcephaly-increase as "Zika-microcephaly" in October 2015, there were close to zero confirmed proven cases of Zika. That undercuts the first sentence. How could Zika have been *"found"* to be circulating in Brazil without having been found to be circulating in Brazil?

Dr. Brito, credited with connecting one unproven event (Zika) with another (microcephaly) makes a similar leap in this quoted comment, with a bit of circularity in logic. He is saying that once the Zika epidemic *"has been confirmed, diagnosis is clinical"* with no place for laboratory meddling. This makes very little sense. The laboratory would be essential in confirming a Zika-diagnosis that was so completely brand-new as never to have been seen clinically at all, anywhere on earth.

This was the case with Covid-19. Early cases involved clinical diagnosis, but ultimately laboratory confirmations were needed to establish the diagnosis broadly, and differentiate it from other illnesses with similar presentation.

Why wouldn't that have been the case with Zika? Early difficulties in finding appropriate and adequate laboratory testing for Zika would have been an issue, no doubt, but once things ramped up, Zika testing would have been crucial– clearly the opposite of Dr. Brito's statement.

Research physicians have some advantages here. They are at a remove from the personal and emotional situation of a tragic birth. They are able to compile and view, through the lens of time and geography, changes and trends that individual doctors and clinicians cannot. They have access to more-refined laboratory-reagents and worldwide - networked information-technology and -databases. Additionally, they can specialize in one particular area, repeatedly researching one question—an advantage that clinicians who continue to juggle multiple other ongoing patients and business-concerns cannot have.

Regardless, the news that Zika caused birth defects spread rapidly. The alarm sounded, and people panicked.

Brazilians don't conduct regular fire drills to the same extent as elsewhere.[232] The notion of a "false alarm" may not be embedded in their cultural consciousness in the same way that it is in North America and Europe. Through media and word of mouth, fears of a Zika-outbreak spread far faster than cautionary, evidence-based science. [233]

> *"In the case of the Zika epidemic, ... the media could not initially find answers ... from researchers and scientific institutions.... Although the emergency (later) spurred research, it may take months (or years) ... to obtain enough knowledge to understand the situation."* Profs. Simone Evangelista Cunha[234] and Marcelo Garcia, 2018

As mentioned in earlier chapters, people transformed their lives. Interim Zika-fear prevented hundreds of thousands of births.[235] Since there was no effective test for Zika at the time, people took every precaution they thought necessary. Brazil's military was mobilized, the WHO declared an emergency; travel restrictions were set; property and personal rights fell to forced insecticide spraying. These lifestyle changes presaged the gauntlet of lockdowns, quarantines, travel bans, business- and school- shutdowns, job losses, disruptions, and social distancing enacted and endured during the *"Age of COVID"*.

Without drawing any false equivalency between the two situations -- nonetheless, the impact of Zika on the world economy was substantial. Billions of dollars in wealth and business opportunities were lost, predominantly in Brazil.[236] The measures, restrictions, and fears surrounding Zika became a tornado of activity: larger and more intimidating than warranted. The question—quite simply—is "why?"

Perhaps contributing to the absence of a reconsideration of the Zika pandemic is its particular geography. Distant problems (in this case Brazil's) are less pressing to resolve for the "medical superpowers" (of East Asia, Europe, North America). And while the issue still looms in Brazil, it has been overshadowed by other pressing problems: dire economic straits, social issues, and more recently Covid. As a result, in Brazil, Zika seems largely forgotten, as seen through Google Trends. [237]

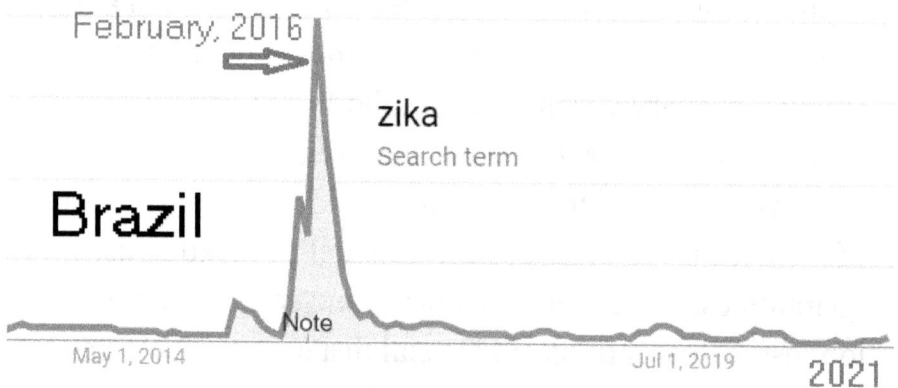

Admittedly, people avoid unpleasant memories, including or especially those of a pandemic. Many people, I'm sure, would love (as soon as possible) to be able to forget

COVID and move on with their lives. Likewise, Brazilians (outside of mothers of affected children)[238] have had no particular desire to remember Zika after the initial wave of panic died down. The microcephalics' mothers can't forget, and they're unlikely to have their minds changed about causality, with the current Brazilian federal stipend for Zika-microcephaly cases, set up as a special fund.[239]

However, the conversation surrounding Zika is not finished. We, as a global community, need to understand the impact Zika had not only during its most fearful immediacy in 2016 (with worries rampant of ruined pregnancies, ruined young lives) but also during the utter "Zika-quiet" in the years thereafter. Aside from WHO's and CDC's eventually quietly downgrading the "emergency",[240] without retracting the premise of the underlying danger -- there has been no positive or hopeful mention of Zika to the general population equivalent in magnitude to the initial panic: not from medicine or science, not from the news media, nor from governments and agencies. Each set of Zika-experts should have been ecstatic that these predictions hadn't come true... yet ... "crickets".[241]

In fact, to the extent that pronouncements have been made, they all intend to keep the fear elevated, rather than bring reassurance that the initial claims might have been exaggerated. For instance, in November 2016 with Zika/microcephaly claims' having essentially disappeared,

WHO's Health Emergencies' Dr. Peter Salama insisted on *"not downgrading the importance of Zika"*; reassuring (sic) that Zika and its WHO's response were both *"here to stay, in a very robust manner."*[242]

In fact, any of the sporadic and quiet mentions of Zika's disappearance were accompanied by either a sense of regret (!) that it could no longer serve as the anticipated tool for "social action" (i.e. overturning Brazil's abortion restrictions; or bringing foreign-aid money to Brazil's poor Northeast)[243] -- or a doubling down on the *"science"*: that Zika's quiet phase presaged a future surprise return, much as in horror movies. These responses coincided with the continued push to pour money into Zika-research even if it is no longer around.[244]

Good news is like a magic elixir, bringing people joy. Experts more invested in the people's well-being than their own reputations or stubborn sense of infallibility would be jumping over each other to be the first to announce Zika's return to medical irrelevancy. A populace previously brought to fear, to dread, and to panic are owed the relaxation, the calm, the assuring presence—or in this case the assuring absence—of one fewer worrisome disease.

Announcements of ***"Zika is gone! Rejoice!"*** never occurred, Zika was later found in other global locations as well as in Brazil during subsequent years—yet nowhere,

even when examined with greater scrutiny, was it causing the predicted microcephaly increases.[245]

People's lives were turned upside down in fear in the Zika pandemic. If Zika was a false alarm, what can we learn from it to avoid similar false alarms in the future? Too many emergencies, too many alarms can make us begin to ignore these warnings. *"Alarm-fatigue"*[246] undercuts the needed reactions when a real threat emerges.[247]

We can't retrieve the billions spent investigating Zika, but "at the end of the day" very little evidence of this connection has been found. According even to Brazil's own Secretariat of Health Surveillance, in 2017: [248]

> *In view of the apparent resurgence of ZIKA infection ... early in 2016, we anticipated a further increase in cases of microcephaly later in the year. But such a resurgence did not happen* [249]

The news media holds some responsibility. "If it bleeds, it leads" is a phrase often attached to the news-generating industry, the media. Newsworthiness, the determination of which stories get top billing, comes down to a variety of factors – but undoubtedly the more lurid ones rise to the top: that means violence, conflict, destruction, fear, mystery. "Shocking" news attracts and keeps viewers, the better with which to keep advertisers happy and the news-business afloat. Retractions show fallibility and undercut the sense of reliability of the newscasters themselves.[250] [251] Humility doesn't bring dividends for journalists (or scientists).

According to Adam Marcus of Retraction Watch,[252] a news outlet is under no legal obligation to revisit the original piece. That said, when asked if he thought a magazine had a duty to issue a clarification or retraction under such circumstances, he said, *"Ethically, I think they should consider following up on their original reporting."*

In the meanwhile, the constant repetition of the "factoid"[253] has brought about its own "illusion of truth".[254] The human mind individually and societally attaches to that which is "the gospel truth" partly through its having been heard over and over again.

It's doubtful that anyone in Brazil from research scientists to the barrios' mothers affected by the Zika-scare have ever seen or heard of this absence of a real microcephaly burst (2015). That's a shame. Mothers still in Brazil and around the tropics worry over Zika. In the US, Zika-testing (and resultant fears) persist.

Overturning Zika doesn't recommend a science version of "*The International Court of Justice* (at The Hague)", because invariably it would be politicized. Ideally "fresh air is the best disinfectant": healthy debate's equilibrium coincides with "the truth", more reliably than any other

method. People in the professions, whether the media, science, politics, or medicine need to be just that, "professional"– and aim for proper deliberation – along with periodic reviews to make sure that the foundational ideas within any given subject are valid.

Historically, the best scientists have been those ***doubting*** the underlying science. They haven't always had an easy time of it: from Einstein's being ridiculed[255] to Galileo's being put on trial and forced to recant under threat of death.

Although open debate and scientific dialogue are still possible– and one remains hopeful there's enough independence in science, in enough places on earth, and within enough hearts and minds to allow different theories' emergence -- recent events have brought similar skepticism. A fair amount of scientific debate is quashed under finger-pointing claims of "misinformation". "*Shut up*" is not an adequate debate point but is used quite effectively by those in power to control the narrative. There are reasons to remain hopeful: scientists (one presumes) all begin as neutral and natural skeptics: questioning, experimenting and checking and following where the results may lead.

Overturning Zika's Entirely Different "_Mission_":
Relief Through Knowledge

Overturning Zika required a year of thorough and entirely unpaid research. This was an endeavor sparked by curiosity, taken up as devotion, finished as a challenge. *Overturning Zika* calls into question each of the three supports upon which the Zika-microcephaly theory rests – only one of which needs to be shaky for the entirety to be re-examined – or, if proved false, for the Zika-microcephaly connection to be invalidated.

The Zika-microcephaly axiom, created in a rush, unverified at the time, instantly brought a reverberating fear that enlarged it, through cases flooding the site of the original claims -- an aftereffect of understandable maternal overcaution and sympathetic medical overdiagnosis. This phenomenon short-circuited early active scientific doubts and auto-generated "Titanic Syndrome"; a panic that replaced orderly deliberation with emergency actions.

The actual Titanic's path, crash, and tragedy had no underlying conspiracy. Staff and passengers shared goals: saving lives; women and children first. One can assume similarly of all the major actors in this intense year-long Zika-microcephaly drama: the best intentions, and wanting to do the most good, quickly, efficiently. That's not to say that either situation produced the best possible outcome– although the Titanic's leaders had no real choices: no time

for reconsideration; no ability to redirect the inevitable tragedy.

Those navigating the Zika-microcephaly theory have had far more opportunity for course-correction. Those attempting to do so have been few in number. Even now, more than five years into the ongoing, fortunate failure of the Zika-microcephaly connection theory – there's a glaring absence of those courageous enough to question the original premise. As Zika rose in prominence so too did certain careers. Similarly, Zika-generated research-grant funding creates a dependency that recipients would rather not jeopardize. The heartfelt Brazilian Zika-microcephaly subsidies to unfortunate families have further entrenched the syndrome.

There are, obviously, some parallels with today's "great pandemic", Covid 19– although both SARS and its reawakening, Covid-19, come closer to microbiology's gold standard of proving infectiousness, reproducibility of results, the "Koch-postulates"[256] than Zika does: following *"if this, then that"* -logic.

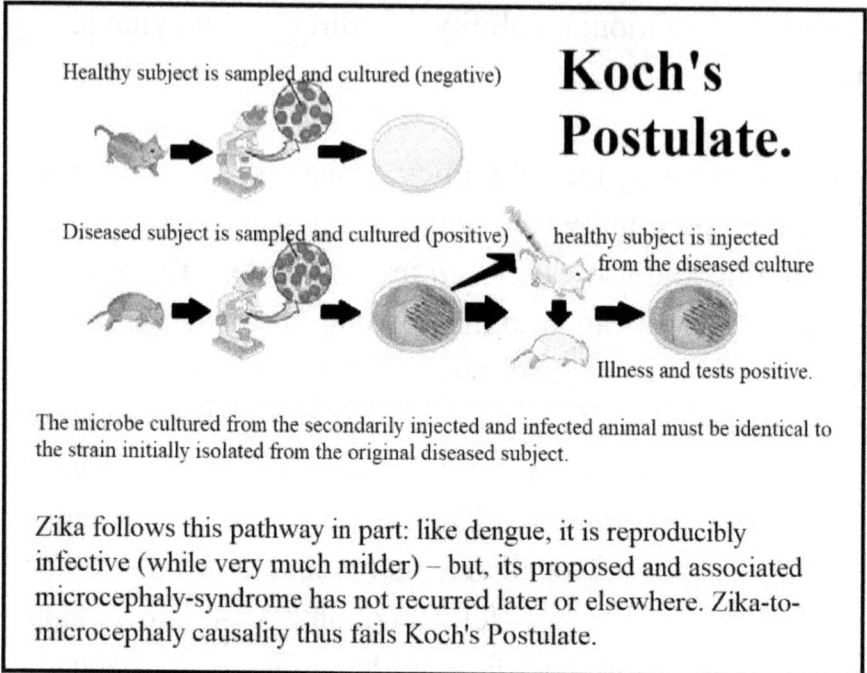

Healthy subject is sampled and cultured (negative)

Koch's Postulate.

Diseased subject is sampled and cultured (positive)   healthy subject is injected from the diseased culture

Illness and tests positive.

The microbe cultured from the secondarily injected and infected animal must be identical to the strain initially isolated from the original diseased subject.

Zika follows this pathway in part: like dengue, it is reproducibly infective (while very much milder) – but, its proposed and associated microcephaly-syndrome has not recurred later or elsewhere. Zika-to-microcephaly causality thus fails Koch's Postulate.

Introduced into the proper group, say pre-vaccine nursing-home resident 85-year-olds, SARS-2 will produce and reproduce its damage almost across the board– no matter the country. Conversely, Zika had NEVER caused human disease anywhere on earth prior to 2015's conjectured reversal of "mild dengue" -cases in Brazil – done, of course, without laboratory testing. And in the aftermath of the whirlwind of relabeling mild dengue as Zika, the following year, Zika fizzled in Brazil, and fizzled everywhere else on earth: not bringing illness directly on infection, and even more fortunately: not resulting in notable increases of congenital birth defects, principally microcephaly.

"The Scientific Method" has four steps for developing hypotheses into theories, theories into scientific "law".

1. Observe and describe the phenomenon.
2. Create a hypothesis that explains the phenomenon.
3. Use this hypothesis to attempt to predict other related phenomena or the results of another set of observations.
4. Test the performance of these predictions using independent experiments.[257]

THE SCIENTIFIC METHOD
HYPOTHESES, MODELS , THEORIES AND LAW

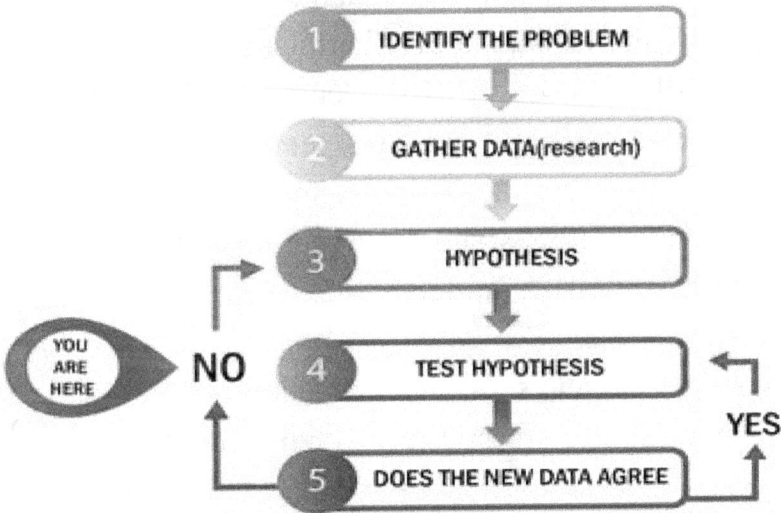

It is unethical and inhumane to test dangerous infections on humans, so step four can take some time and careful planning to create the proper circumstances for evaluating the hypothesis. Instead, this usually plays out in the natural experimentation between different groups of people. In the case of Zika-microcephaly the test-experiments occurred as Zika (it is assumed) traveled to other countries, or recurred in Brazil the subsequent year.

Once 2015 northeast Brazil's Zika-microcephaly phenomenon did not recur in Brazil or anywhere else the subsequent year, the answer to #5, **"Does the New Data Agree?"** became "NO". **The Scientific Method** requires at that point either reworking or abandoning the original hypothesis.

In brief, dengue is endemic in Brazil and is essentially the biological twin of Zika (forever harmless, never active in humans until supposedly 2015). Dengue has a million or more Brazilian symptomatic cases per year and is carried by precisely the same mosquito, *Aedes aegypti*, tagged also as Zika's vector. In relabeling (without laboratory confirmation, at all) certain mild dengue cases, Zika became in a sense, *"The People's Choice Award"*[258] amongst intrepid and semi-independent physicians and researchers slightly off the main Brazilian academic grid.

Professional opportunities for advancement coincide with "naming rights" and the discovery of previously unknown illnesses and syndromes. This was not the precise equivalent of the "cold fusion" debacle, but possibly there are similarities, within Zika-microcephaly's set of physicians, researchers, and newsmakers who together

pushed a narrative not necessarily cohesively – but in accord with their own individual or small-group interests.

The very name of the physicians' group (**CHIKV, The Mission)** -- the precise and essential conduit between the two central events: Zika in the spring, microcephaly in the fall; resulting in its member Dr. Brito's creating the Zika-microcephaly theory -- couldn't have been clearer about this: making their search (begun months before any thought of reconfiguring dengue as Zika) for a virion new to Brazil one of heroism and daring, potentially a spiritual mission. The group's movie-title inspiration Jesuits promoted a profit-sharing, anticolonial utopianism. **Overturning Zika** suggests these central theorists' search for a new virion was a placeholder for the method and means to cure certain social ills. Incredibly enough, Zika microcephaly's wave carried with it the world's attention and financial impact at least partially to do so. This is the fascinating back-story of Zika.

Completely separately, later in the year (fall, 2015; Recife, Brazil) a microcephaly epidemic was declared, without any underlying comparison data or federal registry from which to prove (or disprove) the declared increase. Occurring, principally as it did in a poor area of Brazil's poorer North, a "conspiracy of interests"[259] (including normal medical caution, sympathy for the poor babies and their mothers – and a fair amount of grievance-politics, mixed in with media sensationalism) carried the story to its apogee, especially once it was tied together with the lesser-known but also novel and thus mysterious Zika virus.

The Zika-microcephaly phenomenon consists of three separate legs of a stool, all of which have to be completely solid and well-formed for the entirety to stand.

1. Mild dengue cases constituted some new and different illness. Dengue's biologic twin, Zika, which had never been a human illness and for which there was no testing available in Brazil at the time, was that new illness.

2. Recife neuropediatricians' unofficial counts and feelings there was more microcephaly were fact, even without corroborating comparison data on hand— let alone no consistent metric for measurement and diagnosis of the syndrome.

3. Microcephaly, overall very rare, and with multiple causes, none of which seem specific, and almost all of which track with poverty necessarily had to have a new cause for this unverified increase in cases, and that new cause was to be Zika, whose twin dengue had never caused any such syndrome despite millions of Brazilian cases.

None of these three seemed scientifically solid, even at the time. There had been a fair amount of skepticism and counter-argument to each of the three separate novel formulations, but these were no match for the whirlwind new cycle.

The Emperor's New Clothes -paradigm has taken hold in the meanwhile. There is an enormous amount of scientific and financial ballast's shoring up the status quo – and essentially zero questioning from within the medical or scientific academy over the original precepts, despite none of the original dire predictions' coming to pass, e.g. 1 million excess Zika -related microcephaly births per year, worldwide, principally in the tropics, in perpetuity!

Women worldwide still operate under an intense fear that going outside, leaving oneself open to a random mosquito bite, may irrevocably damage the new life within. Always, there are enough reasons for caution and guilt, but Zika (more specifically Zika-microcephaly) is not one of them, nor should it ever have been.

People worldwide (including 3.3 billion in the tropics) "deserve a break today".[260] ***Overturning Zika's*** own **MISSION** is to provide relief, through knowledge. Not every pandemic should stay (indefinitely). Microcephaly has been around as long as humans and other animals have themselves been. There just wasn't any increase in Brazil 2015, there hasn't been any since then in Brazil or worldwide, with or without Zika. Retractions should be made.

These endless Zika pandemic directives and cautions don't come without a cost, and not just in all the research money diverted towards a phenomenon without foundation. More than 100,000 births were avoided in Brazil alone, and likely the same or larger number of families were similarly indirectly affected worldwide.

At a time when we are told not to "question science", it's important to think about and determine what actually is "science". Science has two underlying definitions, both a set of certain knowledge and facts – and as well the process by which those are determined, and thereafter presumably continually questioned. Questioning science is science. *Overturning Zika* hopes to serve as a demonstration and reminder of both these definitions.

The Zika microcephaly theory within the scientific academy may be at the precipice and shortly just beyond – a stage well-known to one Wile E. Coyote, who will not fall until he realizes there is no support for his position. **Overturning Zika** says to hundreds of millions in the tropics seemingly still under threat and in continued fear of the risk of Zika-microcephaly, *"Things are looking up!"* -- and to the somewhat recalcitrant Zika-research community, *"Don't Look Down!"*[261]

INDEX

A

academy, 29, 48, 67, 251, 252
Aedes Aegypti mosquito, 22, 27, 51, 91,
    101, 114, 127, 128, 129, 130, 134,
    135, 138, 157, 198, 202, 248
Aedes Albopictus mosquito, 129, 130
alarm, 22, 76, 117, 120, 181, 236, 241
Alarm, 100, 237, 241
alcohol, 108, 147
arbovirus, 40, 43, 51
Asia, 50, 208, 238
Asia-strain, 50
awareness, 29, 30, 117, 120, 126, 156

B

Bahia, Brazil, 58, 63, 67
Bancel, Stephane, 231
bias, 85, 106, 117, 119, 120
birth, 26, 39, 40, 41, 42, 43, 73, 77, 88,
    101, 105, 106, 109, 117, 119, 132,
    142, 148, 153, 165, 174, 176, 191,
    200, 201, 203, 207, 211, 213, 214,
    233, 236, 246
Bock, Randall S., 2, 4, 5, 7, 8, 9, 10, 13
Brady, Oliver, 192, 197
Brasília, Brazil, 151, 172
Brito, Carlos, 24, 25, 26, 56, 67, 72, 76,
    77, 78, 79, 84, 97, 106, 180, 199,
    234, 235, 236, 249
business, 26, 36, 73, 178, 179, 218, 227,
    236, 237, 238, 241

C

Calendar year 2015, 3, 13, 17, 21, 24,
    25, 26, 30, 31, 41, 42, 43, 51, 53, 56,
    57, 58, 59, 68, 70, 76, 80, 82, 83, 90,
    92, 93, 104, 105, 106, 110, 112, 115,
    117, 118, 119, 120, 125, 126, 134,
    135, 136, 139, 144, 148, 149, 155,
    159, 165, 168, 179, 181, 182, 191,
    192, 197, 199, 202, 210, 211, 216,
    217, 235, 242, 246, 248, 249, 251
Calendar year 2016, 2, 17, 19, 20, 26,
    29, 33, 34, 37, 49, 69, 90, 92, 105,
    117, 120, 122, 126, 129, 132, 133,
    137, 139, 151, 159, 165, 166, 168,
    170, 189, 199, 200, 202, 207, 210,
    217, 220, 222, 224, 230, 239, 241
Calendar year 2017, 2, 34, 90, 91, 161,
    165, 166, 179, 192, 199, 200, 221,
    223, 229, 241
Calendar year 2018, 34, 90, 91, 145,
    155, 199, 223, 225, 229, 237
Calendar year 2019, 21, 51, 90, 92, 166,
    189, 191, 201, 205, 221, 230
Calendar year 2020, 14, 22, 92, 198,
    218, 260
Calendar year 2021, 2, 92, 207, 231, 260
Calendar year 2022, 9, 14, 230, 260
California, 173
Camaçari, Brazil, 58, 62, 64
Campos, Gubio Soares, 24, 58, 96
Campos, Silvia Sardi, 24, 58
Camurça, Silvia, 173
Catholic, 28, 37, 123, 172, 175, 178
CDC, the Centers for Disease Control
    (US), 19, 24, 29, 31, 36, 41, 43, 48,
    49, 85, 104, 132, 184, 206, 207, 223,
    239
chikungunya, 56, 58, 75, 79, 165, 183,
    199, 202
Chikungunya virus, 56, 58, 62, 75, 77,
    79, 80, 249

CHIKV, The Mission, physician group in pursuit of new viruses, 56, 58, 75, 79, 80, 249

Chioro, Arthur, 182

clinical, 6, 21, 24, 43, 48, 56, 58, 59, 67, 83, 85, 110, 164, 180, 185, 209, 214, 223, 235, 236

cofactors, 191, 192, 212

cold fusion, 95, 96, 248

Colombia, 102, 138, 139, 198, 211, 214, 217

compromised, 32, 97, 107

congenital, 17, 18, 25, 39, 40, 69, 70, 71, 82, 88, 137, 141, 142, 144, 151, 164, 185, 191, 192, 207, 210, 213, 214, 246

Congress, US, 186, 230

conjecture, 12, 166, 233

Covid-19, 6, 11, 12, 22, 23, 98, 189, 204, 206, 216, 217, 218, 219, 231, 236, 238, 245

crime, 188, 192

criteria, 67, 80, 84, 85, 119, 120, 131, 159, 161, 168

Cunha, Simone Evangelista, 237

cytomegalovirus, 27, 75, 108, 142, 147

D

data, 4, 6, 25, 29, 31, 32, 39, 80, 87, 90, 93, 100, 114, 115, 118, 119, 121, 132, 133, 142, 144, 155, 159, 164, 166, 168, 178, 191, 193, 197, 202, 203, 233, 234, 248, 249, 250

debate, 4, 31, 94, 100, 186, 233, 242, 243

Dengue, 13, 17, 23, 24, 27, 30, 40, 43, 48, 49, 51, 56, 58, 62, 63, 64, 67, 71, 78, 79, 82, 83, 84, 88, 90, 91, 93, 96, 122, 126, 128, 129, 131, 157, 161, 162, 163, 164, 165, 166, 168, 169, 170, 183, 184, 193, 200, 202, 203,

207, 208, 209, 213, 214, 220, 246, 248, 249, 250

diagnosis, 43, 48, 58, 63, 73, 80, 109, 113, 118, 131, 161, 235, 236, 250

diagnostic, 21, 25, 80, 109, 110, 112, 119, 164, 185, 209, 233

Diniz, Debora, 73, 173

disappearance, 22, 34, 187, 191, 192, 217, 219, 223, 229, 240

disease, 18, 20, 33, 35, 39, 40, 42, 49, 56, 58, 62, 63, 67, 70, 90, 104, 110, 118, 142, 162, 165, 177, 182, 183, 186, 217, 221, 230, 240, 246

Dos Santos, Claudia Nunes Duarte, 66, 67

Dr. Seuss (Theodor Geisel), children's author), 100, 187

Durbin, Anna, 228, 230

Dye, Christopher, 165

E

Einstein, Albert, 30, 54

El Salvador, 217

emergency, 6, 20, 22, 28, 37, 41, 81, 90, 117, 126, 132, 179, 184, 222, 225, 237, 239, 244, 260

ethnic, 28, 151, 153

Europe, 28, 131, 151, 177, 211, 237, 238

evidence, 5, 6, 18, 40, 71, 112, 134, 192, 201, 237, 241

F

false alarm, 100, 237, 241

Fauci, Anthony, 34, 36, 221, 222, 223, 229, 230

fear, 3, 7, 12, 13, 17, 34, 36, 49, 88, 122, 166, 199, 206, 217, 218, 225, 237, 239, 240, 241, 244, 251, 252

FIOCRUZ, Fundação Oswaldo Cruz, Brazil, 66, 91, 92, 93

Flavivirus, 40, 41, 42, 71, 164, 207, 220

Florida, 133

G

Galileo, 243
Garcia, Marcelo, 237
Garrett, Laurie, 175, 176
genetic, 26, 70, 109, 110, 153, 177, 231
genomics, 24, 43, 49, 219
government, 6, 33, 67, 98, 173, 175,
    179, 186, 217, 218, 219, 231, 239

H

Hepatitis C, 40
Human Challenge Trial (a.k.a.
    Controlled Human Infection Model),
    224, 229, 230
hypothesis, 5, 32, 78, 90, 102, 122, 159,
    160, 164, 165, 168, 200, 247, 248
hysteria, 4, 5, 13, 133, 177, 197

I

illness, 18, 20, 24, 27, 28, 39, 41, 43, 53,
    56, 57, 59, 63, 64, 65, 82, 88, 93, 96,
    104, 131, 141, 174, 205, 217, 235,
    246, 250
immunity, 23, 30, 37, 39, 90, 141, 199,
    200, 208, 209, 213
India, 34, 199, 201
Indio Amazonian ethnicity, Brazil, 28,
    151, 152
insect repellent, 33, 123, 159
insecticide, 159, 201, 237

J

Jamaica, 217
journalism, 3, 14, 18, 32, 105, 112, 118,
    126, 133, 166, 170, 177, 179, 180,
    185, 215, 233, 237, 239, 241, 243,
    250, 260

K

Kallás, Esper, 227, 228
Killian, Caitlin, 174
Ko, Albert, 165, 166, 200, 201

L

laboratory, 24, 26, 27, 43, 49, 58, 59, 66,
    76, 80, 101, 105, 106, 109, 126, 161,
    162, 168, 235, 236, 246, 248
Lambrechts, Louis, 198
Latin America, 126, 139, 175, 176, 178,
    198, 214, 223
Latin American, 175, 176, 178, 198,
    223
leak, 79, 206
Luz, Kleber, 24, 56, 58, 65, 66, 67, 77,
    78, 180

M

Marcus, Adam, 242
Marques, Ernesto, 166
McNeil, Donald, 175, 178, 183
medical, 2, 3, 4, 6, 13, 14, 18, 22, 25, 34,
    43, 48, 53, 55, 57, 59, 62, 65, 69, 79,
    85, 87, 93, 101, 141, 179, 185, 225,
    238, 240, 244, 249, 251
Menezes, Jucielle, 77
Mestizo, combined European and
    Indigenous ancestry, 151, 152
metrics, 126, 197, 233
microcephaly, 3, 7, 12, 13, 17, 18, 21,
    22, 25, 26, 27, 28, 29, 30, 32, 33, 34,
    35, 37, 39, 40, 42, 43, 48, 53, 54, 55,
    59, 69, 70, 71, 73, 75, 76, 77, 78, 79,
    80, 81, 82, 83, 84, 85, 87, 88, 89, 90,
    91, 92, 93, 95, 96, 101, 104, 105,
    106, 107, 108, 109, 110, 112, 113,
    114, 115, 117, 118, 119, 120, 121,
    122, 123, 125, 126, 130, 131, 133,
    134, 135, 137, 138, 139, 140, 144,
    146, 148, 149, 151, 152, 154, 155,

156, 157, 159, 160, 161, 162, 164,
165, 166, 168, 169, 170, 172, 173,
174, 176, 177, 178, 179, 180, 185,
189, 190, 191, 192, 193, 196, 197,
198, 199, 200, 201, 202, 203, 205,
207, 209, 210, 211, 212, 214, 215,
216, 217, 221, 223, 225, 230, 231,
233, 234, 235, 239, 241, 242, 244,
245, 246, 248, 249, 250, 251, 252,
260

Moderna pharmaceutical company, 223,
231

mosquito, 12, 17, 20, 22, 27, 33, 40, 49,
51, 91, 101, 114, 125, 127, 128, 131,
134, 135, 147, 150, 151, 159, 164,
198, 201, 209, 213, 231, 234, 248,
251

mothers, 7, 17, 26, 29, 31, 39, 69, 73,
75, 78, 80, 83, 85, 105, 106, 109,
117, 120, 142, 162, 163, 164, 200,
207, 210, 233, 239, 242, 249

mutant, 30, 90, 200, 203, 214

N

neuropediatrician, 25, 70, 75, 78, 80, 83,
115, 250

New York Times, The, 166, 181, 185

Nielson, Karin, 166

O

Olympics, 26, 33, 37

Overturning Zika, (this book, self-
referenced), 3, 4, 5, 7, 8, 9, 11, 12, 13,
18, 23, 32, 36, 54, 93, 124, 133, 144,
176, 190, 192, 196, 197, 204, 206,
207, 208, 209, 210, 211, 212, 214,
215, 216, 231, 242, 244, 249, 251,
252, 260

P

Pandemic, 2, 3, 5, 6, 7, 8, 9, 11, 12, 17,
20, 30, 33, 34, 35, 36, 37, 41, 42, 43,
51, 53, 80, 90, 91, 106, 118, 126,
162, 168, 177, 189, 192, 205, 206,
215, 218, 223, 225, 230, 238, 241,
245, 251, 252

panic, 3, 4, 5, 12, 18, 26, 27, 32, 48, 76,
80, 88, 90, 96, 104, 110, 112, 113,
117, 122, 126, 133, 134, 161, 169,
170, 183, 197, 203, 215, 216, 221,
232, 234, 239, 240, 244

patient, 11, 41, 58, 73, 170, 209

Pernambuco, 67, 80, 157, 199

physician, 7, 13, 18, 69, 105, 109, 206

politics, 4, 53, 54, 185, 217, 225, 243,
249

Polynesia, 136

population, 22, 31, 33, 92, 110, 122,
151, 157, 164, 168, 179, 201, 202,
208, 210, 239

poverty, 22, 27, 53, 108, 119, 144, 146,
147, 149, 150, 151, 153, 155, 164,
172, 173, 175, 180, 215, 240, 249,
250

power, 3, 36, 151, 172, 177, 243

pregnancy, 26, 28, 40, 41, 76, 84, 87, 91,
106, 109, 132, 135, 141, 142, 147,
159, 162, 164, 172, 176, 192, 199,
220

press, 25, 26, 59, 65, 78, 79, 85, 94, 95,
96, 97, 104, 117, 166, 179, 233, 234,
260

primers for Zika serologic testing, 24,
51, 59, 63, 64

public health, 3, 5, 6, 14, 17, 19, 22, 30,
33, 35, 37, 41, 91, 104, 126, 164,
165, 175, 178, 179, 217, 218, 220,
231

Puerto Rico, 183, 184, 212, 215

R

Rajasthan, 34, 199
reagents, 51, 59, 101, 236
Recife, 25, 27, 28, 67, 69, 70, 73, 76, 77,
    78, 80, 83, 110, 114, 115, 120, 134,
    144, 148, 149, 150, 151, 155, 157,
    161, 166, 170, 173, 212, 249, 250
recurrence, 89, 231
registry, 21, 25, 115, 118, 121, 132,
    144, 197, 233, 249
reproductive, 20, 175
Research institution(s), 6, 25, 26, 58, 68,
    76, 80, 91, 96
researcher, 12, 18, 22, 24, 25, 30, 59, 62,
    65, 85, 104, 121, 145, 161, 162, 164,
    180, 192, 193, 200, 221, 224, 225,
    234, 237, 248
restrictions, 36, 37, 98, 172, 219, 237,
    238, 240
review, 5, 9, 14, 18, 21, 25, 26, 35, 36,
    39, 59, 63, 77, 85, 94, 100, 101, 104,
    113, 141, 179, 191, 197, 223, 224,
    229, 233, 234, 243
Rousseff, Dilma, 28, 37
rubella, 63, 65, 75, 108, 141, 142, 147,
    174, 211, 213, 214

S

Salama, Peter, 240
São Paulo, Brazil, 28, 151, 172, 227
Science, 3, 4, 7, 12, 13, 14, 17, 18, 22,
    23, 28, 29, 31, 32, 37, 39, 53, 54, 56,
    59, 60, 62, 63, 65, 69, 76, 79, 85, 90,
    93, 94, 95, 98, 99, 100, 101, 102,
    104, 107, 112, 120, 122, 144, 166,
    170, 175, 177, 180, 183, 185, 190,
    203, 220, 225, 226, 228, 231, 232,
    233, 234, 237, 239, 240, 242, 243,
    244, 247, 248, 251, 252
Scientific Method, 18, 32, 53, 62, 85,
    102, 181, 247, 248

SESAB, Secretary of Health of the State
    of Bahia, 63, 64, 96, 104
sewage, standing water, 30, 150, 151,
    201
Shah, Seema, 229
Shakespeare, William, 95
Sherlock Holmes, by A. Conan Doyle,
    188, 189
short stature, 27, 147
skepticism, 20, 39, 80, 164, 182, 243,
    250
Souza, Adélia, 78
standards, 18, 21, 25, 27, 28, 29, 62, 87,
    91, 93, 105, 108, 112, 113, 114, 117,
    151, 161, 162, 164, 170, 207, 210,
    219, 226, 245

T

Tay-Sachs Disease, and its genetic trait,
    177
testing, 24, 29, 32, 43, 49, 51, 59, 67,
    71, 75, 76, 83, 94, 97, 101, 106, 109,
    114, 122, 126, 177, 211, 233, 236,
    242, 246, 250
Testing, 70, 159
Texas, 133
The Brazilian Health Regulatory Agency
    (ANVISA), 230
The Northeast region, (tropical) Brazil,
    27, 29, 58, 108, 114, 136, 137, 138,
    139, 147, 151, 166, 172, 180, 191,
    192, 193, 196, 200, 208, 210, 240,
    248
The Southeast region, (more temperate)
    Brazil, 136, 151, 152, 208
the tropics, 7, 12, 13, 18, 20, 27, 30, 33,
    35, 39, 102, 107, 128, 129, 131, 132,
    134, 135, 137, 138, 151, 175, 207,
    209, 210, 214, 216, 217, 242, 251,
    252
The Yale Global Health Review, 175
the Zika Public Health Emergency of
    International Concern (PHEIC), 126

TORCH-illnesses (acronym), 75, 76, 183, 211, 214
toxoplasmosis, 75, 108, 147

U

U.S. Zika Pregnancy Registry (USZPR), 132
United States, 31, 69, 131, 174, 223
university, 13, 80, 89, 227, 260

V

vaccine, 6, 33, 174, 217, 218, 219, 220, 221, 223, 224, 226, 228, 229, 230, 231, 246
van der Linden Mota, Vanessa, 70, 72, 73, 75, 76, 78, 84, 161
van der Linden, Ana, 73, 77
van der Linden, Helio, 73
variant, 23, 37, 62, 209, 219
viral, referring to viruses, 17, 48, 62, 76, 141, 231

W

wealth, 27, 150
WhatsApp, 56, 58, 75, 76, 79
WHO, the World Health Organization, 31, 81, 117, 126, 131, 161, 165, 206, 207, 237
women, 12, 19, 22, 28, 33, 40, 79, 88, 101, 122, 123, 166, 173, 175, 176, 180, 192, 211, 217, 219, 232, 244, 251

Y

Yap, Micronesia, 43, 49, 56, 67, 128

Z

Zika Forest, 50
Zika The Emerging Epidemic, by Donald McNeil, 28
Zika vaccine, 217, 220, 223, 224, 230, 231
Zika, From the Brazilian Backlands to Global Threat, by Debora Diniz, 73, 234

NOTES, CITATIONS, REFERENCES

- FOREWORD

[1]Dr Nathi Mdladla - Intensive Care - Dr. George Mukhari Academic Hospital and Sefako Makgatho University https://za.linkedin.com/in/dr-nathi-mdladla-47abbb3a

[2]The Team - PANDA. https://www.pandata.org/team/
[3]Zika Virus Response Updates from FDA.   FDA.gov https://www.fda.gov/emergency-preparedness-and-response/mcm-issues/zika-virus-response-updates-fda

[4]Carol Taccetta, MD, FCAP   https://www.linkedin.com/in/carol-taccetta-md-fcap-50a620229/

[5]Steven P. Templeton, PhD. https://medicine.iu.edu/faculty/21147/templeton-steven

- Why read "Overturning Zika"?

[6] "Investigating Zika-Microcephaly's 'Crash' - The American Journal of ...." 10 Feb. 2022, https://www.amjmed.com/article/S0002-9343(22)00108-5/fulltext.

- PREFACE and dedication

[7]"On Russiagate, Durham Indicts the Press Too - WSJ." 5 Nov. 2021, https://www.wsj.com/articles/on-russiagate-durham-indicts-the-press-too-hoax-media-indictment-11636141080.

CHAPTER
1. Introduction: like a comet, Zika burned brightly, then disappeared

[8] "`If You Are Not a Liberal When You Are Young, You Have No Heart ...." https://www.amjmed.com/article/S0002-9343(16)30193-0/pdf.

[9] "Lessons from the Burst Zika Bubble - American Greatness." 25 Mar. 2020, https://amgreatness.com/2020/03/25/lessons-from-the-burst-zika-bubble/.

[10] "Brazil is 'badly losing' the battle against Zika virus, says health minister." 26 Jan. 2016, https://www.theguardian.com/world/2016/jan/26/brazil-zika-virus-health-minister-armed-forces-eradication.

[11] "Zika virus triggers pregnancy delay calls - BBC News." 23 Jan. 2016, https://www.bbc.com/news/world-latin-america-35388842.

[12] "Model-based projections of Zika virus infections in childbearing ...." 25 Jul. 2016, https://www.nature.com/articles/nmicrobiol2016126.

[13] "Microcephaly Prevalence in Infants Born to Zika Virus-Infected Women." 5 Aug. 2017, https://www.ncbi.nlm.nih.gov/pmc/articles/PMC5578104/.
Prevalence of microcephaly 2.3% &

[14] How many are born each year? 140 million https://ourworldindata.org/births-and-deaths

[15] Booming populations in the tropics 40% of world's population in tropical countries.
Thus: Erstwhile estimated prevalence of microcephaly 2.3% * 40% of world's population in tropical countries* 140 million births worldwide = 1.3 million microcephalic births; a low estimate (because birth rate is higher in the tropics). Many estimates were higher both in percentage penetration, and in worldwide spread. If the other Aedes (albopictus) mosquito had carried it to temperate regions, then this predicted result would have been an order of magnitude higher https://news.mongabay.com/2014/07/booming-populations-rising-economies-threatened-biodiversity-the-tropics-will-never-be-the-same/

[16] Fewer babies were born in Brazil amid Zika outbreak, study says https://www.cnn.com/2018/05/29/health/zika-brazil-births-study/index.html

[17] "Zika Virus in the Americas — Yet Another Arbovirus Threat | NEJM." https://www.nejm.org/doi/full/10.1056/nejmp1600297.

[18] "Zika virus strain linked to microcephaly not found in Rajasthan." 3 Nov. 2018, https://www.business-standard.com/article/news-ians/zika-virus-strain-linked-to-microcephaly-not-found-in-rajasthan-118110300652_1.html.

[19] "Pandemic Zika: A Formidable Challenge to Medicine and Public ...." 16 Dec. 2017, https://academic.oup.com/jid/article/216/suppl_10/S857/4753668.

[20] "Zika prompts urgent debate about abortion in Latin America - The ...." 8 Feb. 2016, https://www.washingtonpost.com/world/the_americas/zika-prompts-urgent-debate-about-abortion-in-latin-america/2016/02/07/b4f3a718-cc6b-11e5-b9ab-26591104bb19_story.html.

21 "Brazil's Militarized War on Zika - Global Spheres Journal - UC Santa ...."
https://gsj.global.ucsb.edu/sites/secure.lsit.ucsb.edu.gisp.d7_gs-2/files/sitefiles/Deoliveira.pdf.

22 "The Zika Virus Epidemic in Brazil: From Discovery to Future ... - NCBI." 9
Jan. 2018, https://www.ncbi.nlm.nih.gov/pmc/articles/PMC5800195/.

CHAPTER
2. Zika, A Shock To The System

23 "Zika Virus Outside Africa - PMC - NCBI."
https://www.ncbi.nlm.nih.gov/pmc/articles/PMC2819875/.

24 "Current Zika virus epidemiology and recent epidemics - PubMed." 4 Jul. 2014,
https://pubmed.ncbi.nlm.nih.gov/25001879/.

25 "Genetic and Serologic Properties of Zika Virus Associated with an ...."
https://www.ncbi.nlm.nih.gov/pmc/articles/PMC2600394/.

26 "Flaviviridae - Wikipedia." https://en.wikipedia.org/wiki/Flaviviridae.
27 Flaviviridae https://veteriankey.com/flaviviridae/
28 "Overview of the Flaviviridae With an Emphasis on the Japanese ...." 1 Aug.
2009, https://academic.oup.com/labmed/article/40/8/493/2504889.

29 Hepatitis C in "Hepatitis C in pregnancy | ADC Fetal & Neonatal Edition."
https://fn.bmj.com/content/84/3/f201.full.
 pregnancy https://fn.bmj.com/content/84/3/F201
30 "Hepatitis C - Chapter 4 - 2020 Yellow Book |
Travelers' Health | CDC." https://wwwnc.cdc.gov/travel/yellowbook/2020/travel-related-infectious-diseases/hepatitis-c.

31 Op cit. Overview of the Flaviviridae With an Emphasis on the Japanese
Encephalitis Group Viruses
https://academic.oup.com/labmed/article/40/8/493/2504889

32 "Historical Perspectives on Flavivirus Research | HTML - MDPI."
https://www.mdpi.com/1999-4915/9/5/97/htm.

33 "Interim Guidelines for the Evaluation of Infants Born to Mothers ...." 27 Feb.
2004, https://www.cdc.gov/mmwr/preview/mmwrhtml/mm5307a4.htm.

CHAPTER
3. Zika, dengue: "separated at birth"?

[34]Break-bone fever https://en.wikipedia.org/wiki/Break-bone_fever

[35] "The Antigenic Structure of Zika Virus and Its Relation to Other ...." 8 Feb. 2017, https://journals.asm.org/doi/10.1128/MMBR.00055-16.

[36] "Virus Origami - RockEDU Online." https://rockedu.rockefeller.edu/component/virus-origami/

[37] Therkelsen MD, Klose T, Vago F, Jiang W, Rossmann MG, Kuhn RJ. Flaviviruses have imperfect icosahedral symmetry. *Proc Natl Acad Sci U S A.* 2018;115(45):11608-11612. doi:10.1073/pnas.1809304115

[38] https://en.wikipedia.org/wiki/Truncated_icosahedron

[39] "Molecular Basis of Differential Stability and Temperature Sensitivity ...." 21 May. 2020, https://www.nature.com/articles/s41598-020-65288-3

[40] "Impact of flavivirus vaccine-induced immunity on primary Zika virus ...." 4 Feb. 2020, https://journals.plos.org/plosntds/article?id=10.1371/journal.pntd.0008034

[41] Comparison of conservation of flavivirus E-protein structure and sequence Nature Structural & Molecular Biology in "Structural biology of Zika virus and other flaviviruses - PubMed." 8 Jan. 2018, https://pubmed.ncbi.nlm.nih.gov/29323278/.

[42] Flavivirus maturation leads to the formation of an occupied lipid pocket in the surface glycoproteins NATURE COMMUNICATIONS (2021) 12:1238 https://doi.org/10.1038/s41467-021-21505-9

[43]Distinction without a difference https://en.wikipedia.org/wiki/Distinction_without_a_difference

[44] "Overview of dengue and Zika virus similarity, what can we learn from ...." https://www.ncbi.nlm.nih.gov/pmc/articles/PMC5870307/.

[45] "The history of zika virus - WHO | World Health Organization." 7 Feb. 2016, https://www.who.int/news-room/feature-stories/detail/the-history-of-zika-virus.

[46] "Dengue in Brazil: Past, Present and Future Perspective." https://apps.who.int/iris/bitstream/handle/10665/163881/dbv27p25.pdf?sequence=1

[47] "Dengue and Zika Virus Diagnostic Testing for Patients with a ... - NCBI." 14 Jun. 2019, https://www.ncbi.nlm.nih.gov/pmc/articles/PMC6581290/. Abbreviations: IgM = immunoglobulin M; NAAT = nucleic acid amplification test; PRNT = plaque reduction neutralization test.

CHAPTER
4."Medicine is a social science, ...

[48] "Multiplexed Biomarker Panels Discriminate Zika and Dengue Virus ...."
https://pubmed.ncbi.nlm.nih.gov/29141913/.
[49]Rudolf Virchow - Wikipedia   https://en.wikipedia.org/wiki/Rudolf_Virchow

[50] "Facts and ideas from anywhere - PMC - NCBI."
https://www.ncbi.nlm.nih.gov/pmc/articles/PMC2760177/.

[51] Zika/Dengue Virus Comparison gif - Medical Illustration by John Liebler
https://www.artofthecell.com/zikadengue-comparison-gif/

[52] "PROF. DR. KLEBER GIOVANNI LUZ - Cepclin."
https://cepclin.com.br/doctor/kleber-giovanni-luz/.

[53] "How a Small Team of Doctors Convinced the World to Stop Ignoring ...." 29
Feb. 2016, https://www.newsweek.com/2016/03/11/zika-microcephaly-connection-
brazil-doctors-431427.html.

[54]Zika Virus in Brazil - The SUS response September 2017
http://portalarquivos2.saude.gov.br/images/pdf/2017/setembro/21/zika-virus-in-
brazil-2017.pdf

[55] "Global expansion of chikungunya virus: mapping the 64-year history."
https://www.sciencedirect.com/science/article/pii/S1201971217300899.

[56] "Preliminary Survey in Five Industries of the Camaçari Petrochemical ...."
https://www.sciencedirect.com/science/article/pii/S0013935183710571.

[57] "Zika: From the Brazilian Backlands to Global Threat: Diniz, Debora ...."
https://www.amazon.com/Zika-Brazilian-Backlands-Global-
Threat/dp/1786991586.

[58] "Zika Virus Outbreak, Bahia, Brazil - PubMed."
https://pubmed.ncbi.nlm.nih.gov/26401719/.

[59] "Identificado vírus causador de doença misteriosa em Salvador e RMS." 29 Apr.
2015, http://g1.globo.com/bahia/noticia/2015/04/identificado-virus-causador-de-
doenca-misteriosa-em-salvador-e-rms.html.

[60] Op cit, "Zika: From the Brazilian Backlands to Global Threat." Debora Diniz

CHAPTER

5. "Haste makes waste" vs. "Too little too late"?

[61] "Identificado vírus causador de doença misteriosa em Salvador e RMS." 29 Apr. 2015, http://g1.globo.com/bahia/noticia/2015/04/identificado-virus-causador-de-doenca-misteriosa-em-salvador-e-rms.html.

[62] "Satélite: diagnósticos de casos de Zika na Bahia podem estar errados." 8 May. 2015, https://www.correio24horas.com.br/noticia/nid/satelite-diagnosticos-de-casos-de-zika-na-bahia-podem-estar-errados/.

[63] Zanluca C, Melo VC, Mosimann AL, Santos GI, Santos CN, Luz K. First report of autochthonous transmission of Zika virus in Brazil. Mem Inst Oswaldo Cruz. 2015 Jun;110(4):569-72. doi: 10.1590/0074-02760150192 https://pubmed.ncbi.nlm.nih.gov/26061233/.

[64] Op cit, "Zika: From the Brazilian Backlands to Global Threat." Debora Diniz

[65] Op cit, "Zika: From the Brazilian Backlands to Global Threat." Debora Diniz

[66] "Carlos Alexandre Antunes de Brito Federal University of Pernambuco." https://www.researchgate.net/profile/Carlos-Brito-11.

[67] "Zika/Microcefalia: 'É um novo capítulo na história da medicina' - JC." 29 Nov. 2015, https://jc.ne10.uol.com.br/canal/cidades/saude/noticia/2015/11/29/zikamicrocefalia-e-um-novo-capitulo-na-historia-da-medicina-210170.php.
[68] Op cit, How a Small Team of Doctors Convinced the World to Stop Ignoring Zika.

CHAPTER
6. The Eureka Moment, Microcephaly in Recife

[69] Dr. Vanessa Van der Linden Mota may have overlooked the obvious: twins have higher rates of microcephaly. These following two articles show that birthing twins is associated with smaller intrauterine growth rate, and result in smaller babies are associated with microcephaly (with or without mental retardation) "What is selective intrauterine growth restriction (sIUGR)? - Children's ...." https://www.childrensmn.org/services/care-specialties-departments/fetal-medicine/conditions-and-services/siugr/. /

[70] "Factors associated with small head circumference at birth among ...." 12 Jun. 2010, https://www.ncbi.nlm.nih.gov/pmc/articles/PMC2989671/.

[71] "The Brazilian Doctors Who Sounded the Alarm on Zika and ...." 29 Jan. 2016, https://www.wsj.com/articles/the-brazilian-doctors-who-sounded-alarm-on-zika-and-microcephaly-1454109620.

[72] "First report of autochthonous transmission of Zika virus in Brazil." 9 Jun. 2015, https://pubmed.ncbi.nlm.nih.gov/26061233/.

[73] "Epidemiology of other arthropod-borne flaviviruses infecting humans." https://pubmed.ncbi.nlm.nih.gov/14714437/.

[74]Op. cit. "Zika: From the Brazilian Backlands to Global Threat." Debora Diniz
CHAPTER

7.   The Family Business

[75] "How to assess and support the child with microcephaly." 17 Sep. 2018, https://www.paediatricsandchildhealthjournal.co.uk/article/S1751-7222(18)30163-X/fulltext.

[76] "Brazilian doctor on connecting dots between Zika, birth defects." 8 Feb. 2016, https://www.cbsnews.com/news/zika-virus-brazilian-doctor-vanessa-van-der-linden-explains-connecting-dots-between-microcephaly/. Accessed 31 Mar. 2022

[77] "TORCH screen: MedlinePlus Medical Encyclopedia." 22 Jul. 2020, https://medlineplus.gov/ency/article/003350.htm.

[78]Op cit "Zika: From the Brazilian Backlands to Global Threat." Debora Diniz

[79] "Vanessa Van Der Linden Mota - The Dialogue." https://www.thedialogue.org/vanessa-van-der-linden-mota/.

[80]Op cit Zika virus: Brazilian doctor talks about connecting the dots between microcephaly, the disease

[81] "TORCH Syndrome - NORD (National Organization for Rare Disorders)." https://rarediseases.org/rare-diseases/torch-syndrome/.
[82]Dr. Vanessa Van der Linden Mota – 2014, Gerente médica AACD-PE; Associacao de Assistencia a Crianca Deficiente Association for Disabled Children Assistance https://aacd.org.br/

[83]Dr. Hélio Van der Linden Jr., 42, neuropediatrician https://www.linkedin.com/in/helio-van-der-linden-junior-5a25732a/

[84]Carlos Brito, Adjunct professor. Universidade Federal de Pernambuco. Recife. Brasil. Zika Virus: A New Chapter in the History of Medicine - Semantic ...."

https://pdfs.semanticscholar.org/cb7c/a460e9806c03e099adacf66d1760e02c64fc.pdf.

[85] Op cit Zika: From the Brazilian Backlands to Global Threat Debora Diniz

[86] "Beware overblown Zika estimates. So far, most reports of birth ... - Vox." 2 Feb. 2016, https://www.vox.com/2016/2/2/10898436/confirmed-zika-virus-microcephaly-cases.

CHAPTER
8. Bridging the New Illness to a New Problem

[87] "30 years of fatal dengue cases in Brazil: a review | BMC Public Health." 21 Mar. 2019, https://bmcpublichealth.biomedcentral.com/articles/10.1186/s12889-019-6641-4.

[88] Op cit. How a Small Team of Doctors Convinced the World to Stop Ignoring Zika.

[89] Selection bias. https://en.wikipedia.org/wiki/Selection_bias

[90] Lack of blinding - Catalog of Bias. https://catalogofbias.org/biases/observer-bias/

[91] Observer bias - Catalog of Bias. https://catalogofbias.org/biases/observer-bias/

[92] Recall bias - Catalog of Bias. https://catalogofbias.org/biases/recall-bias/

[93] "Possible Association Between Zika Virus Infection and Microcephaly." 29 Jan. 2016, https://www.cdc.gov/mmwr/volumes/65/wr/mm6503e2.htm.

CHAPTER
9. Intermission

[94] Ministério da Cidadania concede pensão vitalícia a crianças com microcefalia
"Ministry of Citizenship grants lifetime pension to children with microcephaly
Born between 2015 and 2018 who already receive the Continuous Cash Benefit
have a guaranteed special pension" 4 Sep. 2019, http://mds.gov.br/area-de-imprensa/noticias/2019/setembro/ministerio-da-cidadania-concede-pensao-vitalicia-a-criancas-com-microcefalia.

[95] Felipe Dana | Photojournalist https://www.felipedana.com.br/

[96] Here's the math on this one. 00.1 % of these 500 Zika pregnancies would be on average less than one severe microcephaly birth per year. There are ordinarily around 3 million births per year in Brazil. A rate of 00.03 % of all pregnancies

bringing severe microcephaly would yield around 900. So amongst those severe microcephalics, only around one in a thousand (or fewer) would have been attributable to Zika 2017, 2018 (if Zika does this at all. Remember dengue does not. The two are near twins)

CHAPTER
10. Checks and Balances?

[97] "Cold fusion died 25 years ago, but the research lives on." 7 Nov. 2016, https://cen.acs.org/articles/94/i44/Cold-fusion-died-25-years.html.

[98] "Science Journal Suggests Fraud Possible In Texas A&M Cold ...." 15 Jun. 1990, https://apnews.com/article/2814ebed21ecd7c5062d590eff713cb7.

[99] "Cold Fusion Conundrum at Texas A&M - New Energy Times." http://newenergytimes.com/v2/sr/taubes-fraud-depiction/SCIENCE-ColdFusionConundrumAtTexasA&M.pdf.

[100] "The Fifty Year Rehabilitation of the Egg - PMC - NCBI." 21 Oct. 2015, https://www.ncbi.nlm.nih.gov/pmc/articles/PMC4632449/.

[101] "Integrating evidence, politics and society: a methodology for the ...." 17 Apr. 2018, https://doi.org/10.1057/s41599-018-0099-3.
[102] Dr. Seuss The Zax

[103] "Purity - XKCD." https://xkcd.com/435/.

[104] I know it when I see it - Wikipedia https://en.wikipedia.org/wiki/I_know_it_when_I_see_it
\
[105] Scientific theory - Simple English Wikipedia, the free encyclopedia https://simple.wikipedia.org/wiki/Scientific_theory
[106] "Possible Association Between Zika Virus Infection and Microcephaly." 29 Jan. 2016, https://www.cdc.gov/mmwr/volumes/65/wr/mm6503e2.htm.

[107] "Zika and Microcephaly: Jury Still Out | MedPage Today." 1 Feb. 2016, https://www.medpagetoday.com/neurology/generalneurology/55953.

[108] "Population surveillance for microcephaly | The BMJ." https://www.bmj.com/content/354/bmj.i4815.

[109] "Infection-related microcephaly after the 2015 and 2016 Zika virus ...." https://www.thelancet.com/journals/lancet/article/PIIS0140-6736(17)31368-5/fulltext.

[110]Ibid Infection-related microcephaly after the 2015 and 2016 Zika virus outbreaks in Brazil: a surveillance-based analysis.

[111] Op. cit. How a Small Team of Doctors Convinced the World to Stop Ignoring Zika.

[112]" Post hoc ergo propter hoc
https://en.wikipedia.org/wiki/Post_hoc_ergo_propter_hoc

[113]Jonathan Swift (1667-1745). Political Lying. Vol. III. Seventeenth Century.

CHAPTER
11. The Shifting Sands of Shifting Standards

[114] "Facts about Microcephaly | CDC."
https://www.cdc.gov/ncbddd/birthdefects/microcephaly.html.

[115] "Practice Parameter: Evaluation of the child with microcephaly ... - NCBI."
https://www.ncbi.nlm.nih.gov/pmc/articles/PMC2744281/
https://www.aan.com/PressRoom/home/GetDigitalAsset/8479.

[116] "Autosomal recessive primary microcephaly - Genetics - MedlinePlus."
https://medlineplus.gov/genetics/condition/autosomal-recessive-primary-microcephaly/.

[117] "Investigating microcephaly | Archives of Disease in Childhood."
https://adc.bmj.com/content/98/9/707.

[118] "Microcephaly in Brazil: how to interpret reported numbers?."
https://www.thelancet.com/journals/lancet/article/PIIS0140-6736(16)00273-7/fulltext.

[119] "The Epidemic of Zika Virus–Related Microcephaly in Brazil - NCBI."
https://www.ncbi.nlm.nih.gov/pmc/articles/PMC4816003/.

[120] "Possible Association Between Zika Virus Infection and Microcephaly." 29 Jan. 2016, https://www.jstor.org/stable/24856988.

[121] "Statistical estimation for Large Numbers of Rare Events."
https://www.ling.upenn.edu/courses/cogs502/LNRE.html.

[122] "These 4 colors can be the kiss of death when selling your home." 27 Oct. 2018, https://www.marketwatch.com/story/these-4-colors-can-be-the-kiss-of-death-when-selling-your-home-2016-08-23.

CHAPTER
12. The Microcephaly Bubble

[123] "Ministério da Saúde - Governo Federal do Brasil."
https://www.gov.br/saude/pt-br.

[124] "Beware overblown Zika estimates. So far, most reports of birth ... - Vox." 2
Feb. 2016, https://www.vox.com/2016/2/2/10898436/confirmed-zika-virus-
microcephaly-cases.

[125] "Microcephaly in north-east Brazil: a retrospective study on neonates ...." 1
Nov. 2016, https://www.ncbi.nlm.nih.gov/pmc/articles/PMC5096352/.

[126] "Is Zika responsible for Brazil's microcephaly outbreak? - ABC." 8 Feb. 2016,
https://www.abc.net.au/radionational/programs/healthreport/is-zika-responsible-
for-brazils-microcephaly-outbreak/7148978.

[127] Op. Cit. Microcephaly in north-east Brazil: a retrospective study on neonates
born between 2012 and 2015.

[128]Ibid , figure https://www.who.int/bulletin/volumes/94/11/BLT-16-170639-F2.jpg

CHAPTER
13. Mosquitoes Know No Borders... The Net Result?

[129] "Is Zika responsible for Brazil's microcephaly outbreak? - ABC." 8 Feb. 2016,
https://www.abc.net.au/radionational/programs/healthreport/is-zika-responsible-
for-brazils-microcephaly-outbreak/7148978

[130] "Timeline of Emergence of Zika virus in the Americas - PAHO/WHO."
https://www3.paho.org/hq/index.php?option=com_content&view=article&id=1195
9:timeline-of-emergence-of-zika-virus-in-the-americas&Itemid=41711&lang=en.

[131] "Model-based projections of Zika virus infections in childbearing ...." 25 Jul.
2016, https://doi.org/10.1038/nmicrobiol.2016.126.

[132] Washington University School of Medicine. (2020, February 18). Why Zika
virus caused most harmful brain damage to Brazilian newborns: Findings help
explain why microcephaly cases less common in other Zika
outbreaks. *ScienceDaily*. Retrieved April 1, 2022 from
www.sciencedaily.com/releases/2020/02/200218163106.htm

133 "Zika Virus Transmission from French Polynesia to Brazil - PMC - NCBI." https://www.ncbi.nlm.nih.gov/pmc/articles/PMC4593458/.

134 "Zika virus outbreak and the risk to Europe." https://www.euro.who.int/en/health-topics/health-emergencies/zika-virus/zika-virus-outbreak-and-the-risk-to-europe2.

135 "ZIKA - competence of Aedes aegypti and albopictus vector species." http://www.euro.who.int/__data/assets/pdf_file/0007/304459/WEB-news_competence-of-Aedes-aegypti-and-albopictus-vector-species.pdf.

136 "Prevalence charts and tables | EU RD Platform." 3 Jan. 2022, https://eu-rd-platform.jrc.ec.europa.eu/eurocat/eurocat-data/prevalence_en.

137 "Population-Based Surveillance for Birth Defects Potentially Related ...." https://www.cdc.gov/mmwr/volumes/69/wr/mm6903a3.htm.

138 "Zika Research Projects List - PAHO/WHO." https://www3.paho.org/zika-research/index.php?start=680.

139 "Scientists are bewildered by Zika's path across Latin America - The ...." 25 Oct. 2016, https://www.washingtonpost.com/world/the_americas/scientists-are-bewildered-by-zikas-path-across-latin-america/2016/10/25/5e3a992c-9614-11e6-9cae-2a3574e296a6_story.html.

CHAPTER
14. Viral Penetration, Rubella versus Zika

140 "Pregnancy and Rubella | CDC." https://www.cdc.gov/rubella/pregnancy.html.
141 "Consequences of confirmed maternal rubella at successive stages ...." https://pubmed.ncbi.nlm.nih.gov/6126663/.

142 "Management of Pregnancies with Confirmed Cytomegalovirus Fetal ...." 6 Apr. 2013, https://www.karger.com/Article/Fulltext/342752.
Considering any CNS disability (hearing loss, mental retardation, cerebral palsy, seizures, chorioretinitis), 11/34 (32%) first-trimester cytomegalovirus cases were affected.

143 "Effectiveness of prenatal treatment for congenital toxoplasmosis." https://www.thelancet.com/journals/lancet/article/PIIS0140-6736(07)60072-5/fulltext.
The meta-analysis of individual patient data found that crude risk of ocular lesions diagnosed in the first year of life was 14% in the European cohort and 47% in the South American cohorts

[144] "Thalidomide-induced teratogenesis: History and mechanisms - NCBI." 4 Jun. 2015, https://www.ncbi.nlm.nih.gov/pmc/articles/PMC4737249/.

[145] "Possible Association Between Zika Virus Infection and Microcephaly." 29 Jan. 2016, https://www.jstor.org/stable/24856988.

[146]Op. Cit. Microcephaly in north-east Brazil: a retrospective study on neonates born between 2012 and 2015

[147] "Prevalence and Risk Factors for Microcephaly at Birth in Brazil in 2010." https://pediatrics.aappublications.org/content/pediatrics/141/2/e20170589.full.pdf.

CHAPTER
15. A closer look at Brazil's Zika-microcephaly epicenter, Recife Brazil.

[148] "Microcephaly epidemic related to the Zika virus and living conditions ...." 12 Jan. 2018, https://bmcpublichealth.biomedcentral.com/articles/10.1186/s12889-018-5039-z.

[149] "In Brazil the wounds of slavery will not heal | DW | 13.05.2018 - DW." 13 May. 2018, https://www.dw.com/en/in-brazil-the-wounds-of-slavery-will-not-heal/a-43754519.

[150] "Our Favela(s) - passport2freedom.org." http://www.passport2freedom.org/home/home-3/favela/.

[151]Personal communication with Brazilian student.

[152]Op. Cit. The Brazilian Doctors Who Sounded the Alarm on Zika and Microcephaly

[153] "The Brazilian Slum Children Who Are Literally Swimming in Garbage." 30 Jan. 2014, https://www.vice.com/en/article/kwpwja/the-brazilian-slum-children-who-are-literally-swimming-in-garbage-0000197-v21n1.

[154] "Urban Sewage in Brazil: Drivers of and Obstacles to Wastewater ...." https://www.die-gdi.de/uploads/media/DP_26.2016.pdf.

[155]Race and ethnicity in Brazil - Wikipedia https://en.wikipedia.org/wiki/Race_and_ethnicity_in_Brazil#Ethnicities_by_region

[156]Evaluation of the child with microcephaly (an evidence-based review) Stephen Ashwal, MD et al. American Academy of Neurology 06/28/2013
https://www.aan.com/PressRoom/home/GetDigitalAsset/8479

[157]Growth and inequalities of height in Brazil (1939-1981) Monasterio, Leonardo M and Noguerol, Luiz Paulo Munich Personal RePEc Archive 08/01/2005
https://mpra.ub.uni-muenchen.de/769/1/MPRA_paper_769.pdf

[158]Narrowing socioeconomic inequality in child stunting: the Brazilian experience, 1974-2007 Monteiro CA, Benicio MH, Conde WL, Konno S, Lovadino AL, Barros AJ, Victora CG. Bull World Health Organ. 2010 Apr;88(4):305-11. 04/01/2010
https://pubmed.ncbi.nlm.nih.gov/20431795/

[159]Prevalence and Risk Factors for Microcephaly at Birth in Brazil in 2010 Antônio A. Silva, et al. Pediatrics February 2018; 141 (2) 02/01/2018
https://publications.aap.org/pediatrics/article/141/2/e20170589/38041/Prevalence-and-Risk-Factors-for-Microcephaly-at

[160]Pre-Zika Microcephaly in Brazil: Closer to the Elusive Baseline and New Questions Raised Elizabeth Dufort, Jennifer White Pediatrics February 2018; 141 (2): e20173811. 02/01/2018
https://publications.aap.org/pediatrics/article/141/2/e20173811/38051/Pre-Zika-Microcephaly-in-Brazil-Closer-to-the

[161]File:Global Aedes aegypti distribution (e08347).png.
https://en.wikipedia.org/wiki/File:Global_Aedes_aegypti_distribution_(e08347).png

[162]Risk of microcephaly after Zika virus infection in Brazil, 2015 to 2016. Jaenisch, Thomas et al. Bulletin of the World Health Organization vol. 95,3 (2017): 191-198 03/01/2017 https://www.ncbi.nlm.nih.gov/pmc/articles/PMC5328112/

[163]Search term: microcefalia Google trends .
https://trends.google.com/trends/explore?date=2015-10-05%202016-05-05&geo=BR&q=microcefalia

[164] Zika Virus Infection and Associated Neurologic Disorders in Brazil | NEJM. de Oliveira, Wanderson; Christopher Dye, et al. New England Journal of Medicine 03/29/2017 https://www.nejm.org/doi/full/10.1056/nejmc1608612

[165]Perinatal analyses of Zika- and dengue virus-specific neutralizing antibodies: A microcephaly case-control study in an area of high dengue endemicity in Brazil Castanha PMS, et al. Neglected Tropical Diseases, The Public Library of Science 03/11/2019 https://pubmed.ncbi.nlm.nih.gov/30856223/

CHAPTER
   16. Hypothesis Testing: When Results Fail Beliefs

[166]Medical Mystery: Why Did The Number Of New Microcephaly Cases In Brazil Drop In 2016? Michaeleen Doucleff NPR 03/30/2017
https://www.npr.org/sections/goatsandsoda/2017/03/30/521925733/why-didnt-zika-cause-a-surge-in-microcephaly-in-2016

CHAPTER
   17. Never Let a Crisis Go to Waste

[167]never let a serious crisis go to waste.... Rahm Emanuel Quotes. BrainyQuote.com 01/16/2022
https://www.brainyquote.com/citation/quotes/rahm_emanuel_409199

[168]All the "Quotes" Churchill Never Said Richard Lang worth 11/08/2018
https://richardlangworth.com/quotes-churchill-never-said-1

[169]Surge of Zika Virus Has Brazilians Re-examining Strict Abortion Laws (Published 2016) Simon Romero The New York Times 02/03/2016
https://www.nytimes.com/2016/02/04/world/americas/zika-virus-brazil-abortion-laws.html

[170]Opinion | The Zika Virus and Brazilian Women's Right to Choose (Published 2016) Debora Diniz The New York Times 02/08/2016
https://www.nytimes.com/2016/02/08/opinion/the-zika-virus-and-brazilian-womens-right-to-choose.html

[171]The California Therapeutic Abortion Act: An Analysis Brian Pendleton 19 Hastings L.J. 242 11/01/1967
https://repository.uchastings.edu/cgi/viewcontent.cgi?article=1968&context=hastings_law_journal

[172]Could There Be a Silver Lining to Zika? - Caitlin Killian, 2017 Killian, Caitlin. Contexts, vol. 16, no. 1, pp. 36–41 02/01/2017
https://journals.sagepub.com/doi/full/10.1177/1536504217696062

[173]Zika as a Catalyst for Reproductive Rights Reform in Latin America GRACIE JIN yale global health review 05/14/2017
https://yaleglobalhealthreview.com/2017/05/14/zika-as-a-catalyst-for-reproductive-rights-reform-in-latin-america/

[174]The World's Abortion Laws Center for Reproductive Rights .
https://maps.reproductiverights.org/worldabortionlaws

[175]"The Coming Plague: Newly Emerging Diseases in a World Out of Balance
Laurie Garrett   Picador 08/25/2020 https://www.barnesandnoble.com/w/the-coming-plague-laurie-garrett/1003040038"

[176]Review: 'Zika' Tracks the Trajectory of an Epidemic Laurie Garrett The New
York Times 07/28/2016 https://www.nytimes.com/2016/07/29/books/review-zika-tracks-the-trajectory-of-an-epidemic.html

[177]"Zika Virus and the Hypocrisy of Telling Women to Delay Pregnancy
Emma Saloranta Winiecki huffpost 01/29/2016
https://www.huffpost.com/entry/zika-virus-and-the-hypocrisy-of-telling-women-to-delay-pregnancy_b_9090476

[178]The Impact of the Zika Outbreak on Women and Girls in Northeastern Brazil |
HRW Margaret Wurth, et al. Human Rights Watch 07/12/2017
https://www.hrw.org/report/2017/07/13/neglected-and-unprotected/impact-zika-outbreak-women-and-girls-northeastern

[179] "'Big' industry | Latent Paradigm." 13 Sep. 2012,
https://latentparadigm.wordpress.com/2012/09/13/big-industry/.

[180]View of Zika Meg Stalcup Medicine Anthropology Theory 09/10/2018
http://www.mcdanthrotheory.org/article/view/4881/6824

[181]Zika From the Brazilian backlands to global threat; Debora Diniz— Reviewed
by Meg Stalcup
10 Sep 2018 doi.org/10.17157/mat.5.4.62
http://journals.ed.ac.uk/index.php/mat/article/download/4881/6823?inline=1

[182] BOOK AND FILM REVIEWS From the Brazilian backlands to global threat
Reviewed by Meg Stalcup Meg Stalcup Medicine Anthropology Theory
09/10/2018 http://medanthrotheory.org/site/assets/files/11056/mat-5_4_-626-stalcup.pdf

[183]Zika: The Emerging Epidemic: McNeil Jr., Donald G.: 9780393353969. Donald
G. McNeil Jr. W. W. Norton & Company; 1st edition 06/28/2016
https://www.amazon.com/Zika-Emerging-Donald-McNeil-Jr/dp/0393353966

[184]Alarm Spreads in Brazil Over a Virus and a Surge in Malformed Infants
(Published 2015) Simon Romero The New York Times 12/30/2015
https://www.nytimes.com/2015/12/31/world/americas/alarm-spreads-in-brazil-over-a-virus-and-a-surge-in-malformed-infants.html

[185]Arthur Chioro - Wikipedia https://en.wikipedia.org/wiki/Arthur_Chioro

[186] "What Would It Take to Prove the Zika–Microcephaly Link - Scientific ...." 28 Jan. 2016, https://www.scientificamerican.com/article/what-would-it-take-to-prove-the-zika-microcephaly-link

[187]"Ex-Times Reporter Who Used Racial Slur Publishes a Lengthy Defense Marc Tracy The New York Times 03/01/2021 https://www.nytimes.com/2021/03/01/business/donald-mcneil-new-york-times-racial-slur.html

[188]"U.S. Becomes More Vulnerable to Tropical Diseases Like Zika (Published 2016)
Donald G. McNeil Jr. The New York Times 01/04/2016 https://www.nytimes.com/2016/01/05/health/us-becomes-more-vulnerable-to-tropical-diseases-like-zika.html

[189]The Rise and Spread of the Zika Virus collections 07/03/2007 https://static01.nyt.com/packages/pdf/tbooks/rise_and_spread_of_the_zika_virus_n ytimes_102616.pdf?mcid

[190]"Zika Response Funding: Request and Congressional Action Susan B. Epstein, Sarah A. Lister
Congressional Research Service 09/30/2016 https://sgp.fas.org/crs/misc/R44460.pdf

CHAPTER
18. Theories of Zika's Disappearance Abound

[191]The Adventure of Silver Blaze Arthur Conan Doyle The Memoirs of Sherlock Holmes 12/15/1892 https://etc.usf.edu/lit2go/40/the-memoirs-of-sherlock-holmes/573/adventure-1-silver-blaze/

[192]Lessons from Sherlock Holmes: Paying Attention to What Isn't There Maria Konnikova SCIENTIFIC AMERICAN 08/23/2011 https://blogs.scientificamerican.com/guest-blog/lessons-from-sherlock-holmes-paying-attention-to-what-isnt-there

[193]CDC Concludes Zika Causes Microcephaly and Other Birth Defects | CDC Online Newsroom CDC 04/13/2016 https://www.cdc.gov/media/releases/2016/s0413-zika-microcephaly.html

[194]Attempts at Debunking "Fake News" about Epidemics Might Do More Harm Than Good Gary Stix SCIENTIFIC AMERICAN 02/14/2020

https://www.scientificamerican.com/article/attempts-at-debunking-fake-news-about-epidemics-might-do-more-harm-than-good/

[195]The Zika virus epidemic 3 years on: a personal perspective G. Malinger Ultrasound In obstetrics and gynecology 12/18/2018
https://doi.org/10.1002/uog.20199

[196]Zika Virus Infection — After the Pandemic Didier Musso, M.D., Albert I. Ko, M.D., and David Baud, M.D., Ph.D. NEJM 10/10/2019
https://www.nejm.org/doi/full/10.1056/NEJMra1808246

[197]The association between Zika virus infection and microcephaly in Brazil 2015–2017: An observational analysis of over 4 million births Brady OJ, Osgood-Zimmerman A, Kassebaum NJ, Ray SE, de Araújo VEM, et al. PLOS Medicine 16(3): e1002755 03/05/2019
https://journals.plos.org/plosmedicine/article?id=10.1371/journal.pmed.1002755
[198] Ibid., figure 1
[199] Ibid., figure 4

[200]"Enhanced Zika virus susceptibility of globally invasive Aedes aegypti populations
Journal Article, Fabien Aubry, et al. SCIENCE • 20 Nov 2020 • Vol 370, Issue 6519 • pp. 991-996 11/20/2020
https://www.science.org/doi/10.1126/science.abd3663"

[201]Medical Mystery: Why Did The Number Of New Microcephaly Cases In Brazil Drop In 2016? Michaeleen Doucleff NPR 03/30/2017
https://www.npr.org/sections/goatsandsoda/2017/03/30/521925733/why-didnt-zika-cause-a-surge-in-microcephaly-in-2016

[202]"Perinatal analyses of Zika- and dengue virus-specific neutralizing antibodies: A microcephaly case-control study Castanha PMS, et al. Neglected Tropical Diseases, PLOS 03/11/2019 https://pubmed.ncbi.nlm.nih.gov/30856223//"

[203]" A Single Mutation Helps Modern Zika Cause Birth Defects
Dina Fine Maron SCIENTIFIC AMERICAN 09/28/2017
https://www.scientificamerican.com/article/a-single-mutation-helps-modern-zika-cause-birth-defects1/"

[204]"The Neurobiology of Zika Virus: Neuron
Joseph G. Gleeson et al. Neuron, Cell magazine 12/07/2016
https://www.cell.com/neuron/comments/S0896-6273(16)30899-6"

[205]"A Possible Link Between Pyriproxyfen and Microcephaly
Bar-Yam Y. et al. PLoS Curr. 2017 Nov 27;9 11/27/2017
https://pubmed.ncbi.nlm.nih.gov/29362686/"

[206] (except for the Pyriproxyfen theorists who don't so much question the first two conjectures as propose an alternate cause)

CHAPTER
   19. The Experts Respond

[207]"Lyle R. Petersen, MD, MPH | CDC Online Newsroom
Lyle R. Petersen CDC . https://www.cdc.gov/media/spokesperson/sme-bio/petersen-l.html"

[208]"virologist Paolo Zanotto; Arkady Petrov The Rio Times 06/07/2021
https://www.riotimesonline.com/brazil-news/tag/virologist-paolo-zanotto/

CHAPTER
   20. The Broken Promise of the Zika Vaccine; Women's Lives in
       Limbo

[209]Operation Warp Speed: Accelerated COVID-19 Vaccine Development Status and Efforts to Address Manufacturing Challenges | US GAO US Government accountability office https://www.gao.gov/ 02/11/2021
https://www.gao.gov/products/gao-21-319

[210]"Jonas Schmidt-Chanasit MD, in Médico ligado à OMS recomenda que mulheres em área com surto de zika evitem engravidar - BBC News Brasil MD Deputy Director of the WHO Collaborating Centre for Arbovirus and Haemorrhagic Fever bbc 12/04/2015
https://www.bbc.com/portuguese/noticias/2015/12/151203_entrevista_oms_zika_rm"

[211]Ibid
https://www.bbc.com/portuguese/noticias/2015/12/151203_entrevista_oms_zika_rm

[212]Short Answers to Hard Questions About Zika Virus. DONALD G. MCNEIL JR., CATHERINE SAINT LOUIS and NICHOLAS ST. FLEUR New York Times 07/29/2016 https://www.nytimes.com/interactive/2016/health/what-is-zika-virus.html

[213]WHO announces simple, easy-to-say labels for SARS-CoV-2 Variants of Interest and Concern WHO 05/31/2021 https://www.who.int/news/item/31-05-

2021-who-announces-simple-easy-to-say-labels-for-sars-cov-2-variants-of-interest-and-concern

[214]Are we now faced with covid-21? Brian J Ford BMJ 01/06/2022 https://www.bmj.com/content/376/bmj.n3145/rr

[215]"Covid-22 could get worse" Sai Reddy, Danny Schlumpf Blick (Germany) 01/10/2022 https://www.blick.ch/schweiz/eth-forscher-warnt-vor-neuer-corona-super-variante-covid-22-koennte-noch-schlimmer-werden-id16770423.html

[216]"Covid-22 could get worse" Sai Reddy, Danny Schlumpf Blick (Germany) 01/10/2022 https://www.blick.ch/schweiz/eth-forscher-warnt-vor-neuer-corona-super-variante-covid-22-koennte-noch-schlimmer-werden-id16770423.html

[217]Developing Zika Vaccines Nightly Business Report 04/19/2016 https://www.youtube.com/watch?v=EdvFNFsSAUg

[218]Once and future epidemics: Zika virus emerging. GARY J. NABEL AND ELIAS A. ZERHOUNI https://www.science.org/ 03/16/2016 https://www.science.org/doi/10.1126/scitranslmed.aaf4548

[219]Pharma firm issues caution on use of anti-dengue vaccine Tina G. Santos - Reporter / @santostinaINQ Philippine Daily Inquirer / 04:21 PM November 30, 2017 https://technology.inquirer.net/69907/pharma-firm-issues-caution-anti-dengue-vaccine-sanofi-dengvaxia-vaccine-health-dengue

[220]Philippines snubbed advice of experts to tread incrementally in pursuing child dengue immunizations | The Japan Times Japan Times 12/11/2017 https://www.japantimes.co.jp/news/2017/12/11/asia-pacific/science-health-asia-pacific/philippines-snubbed-advice-experts-tread-incrementally-pursuing-child-dengue-immunizations/

[221] Op. cit. Developing Zika Vaccines Nightly Business Report 04/19/2016 https://www.youtube.com/watch?v=EdvFNFsSAUg

[222]Phase 2 Zika Vaccine Trial Begins in US, Central and South America niaid.nih.gov 03/31/2017 https://www.niaid.nih.gov/news-events/phase-2-zika-vaccine-trial-begins-us-central-and-south-america

[223]Scientists begin mid-stage trial of Zika vaccine for first time. Helen Branswell statnews 03/31/2017 https://www.statnews.com/2017/03/31/zika-vaccine-phase-two/

[224]Current status of Zika vaccine development: Zika vaccines advance into clinical evaluation | npj Vaccines Alan D. T. Barrett npj Vaccines 06/11/2018 https://www.nature.com/articles/s41541-018-0061-9#citeas

[225]Tuskegee Study - Timeline - CDC - NCHHSTP. cdc.gov .
https://www.cdc.gov/tuskegee/timeline.htm

[226]Ethical Considerations for Zika Virus Human Challenge Trials Report and Recommendations. niaid.nih.gov 02/10/2017
https://www.niaid.nih.gov/sites/default/files/EthicsZikaHumanChallengeStudiesReport2017.pdf

[227] As massive Zika vaccine trial struggles, researchers revive plan to intentionally infect humans | Science | AAAS. JON COHEN Science 09/12/2018
https://www.science.org/content/article/massive-zika-vaccine-trial-struggles-researchers-revive-plan-intentionally-infect

[228]Ibid

[229]Sunk Cost Fallacy: Know When You Need To Pull The Plug - TechTello Seth Godin Tech Tello . https://www.techtello.com/sunk-cost-fallacy/

[230]Forced to rob cancer research to pay for Zika vaccine push - Anthony Fauci, Jonathan Capehart The Washington Post 09/20/2016
https://www.washingtonpost.com/blogs/post-partisan/wp/2016/09/20/anthony-fauci-forced-to-rob-cancer-research-to-pay-for-zika-vaccine-push/

[231]Moderna Long Term Plan Presentation Stephane Bancel CECP 06/10/2021
https://www.youtube.com/watch?v=oP--FCs-2H4&t=813s

CHAPTER
21. The Verdict

[232]"Factors Influencing Fire Safety in Brazil. Jeremy P. Francisco, Marissa A. Imperiali WPI . https://web.wpi.edu/Pubs/E-project/Available/E-project-031112-202605/unrestricted/Final_Report_-_Mar11.pdf

[233]"Fear time versus science time: discursive disputes over the epidemi... Simone Evangelista Cunha and Marcelo Garcia 06/28/2019
https://journals.openedition.org/cs/750?lang=en"

[234]Simone Evangelista Cunha Postdoctoral fellow in Communication Fluminense Federal University (UFF); Rio de Janeiro state .
https://www.escavador.com/sobre/7307092/simone-evangelista-cunha

[235] Implications of Zika virus and congenital Zika syndrome for the number of live births in Brazil Marcia C. Castro, et al. Proceedings of the National Academy of Sciences 06/12/2018 https://www.pnas.org/content/115/24/6177.long

Forecasts for September 2015 to December 2016 showed that 119,095 fewer births than expected were observed, particularly after April 2016 (a reduction significant at 0.05), demonstrating a link between publicity associated with the ZIKV epidemic and the decline in births.

[236]" How do we know the Zika virus will cost the world $3.5 billion? Jay L. Zagorsky The Conversation 02/23/2016 https://theconversation.com/how-do-we-know-the-zika-virus-will-cost-the-world-3-5-billion-55117"

[237]Google trends, Zika .
https://trends.google.com/trends/explore?date=all&geo=BR&q=zika

[238]"In Brazil, Covid-19 Deals New Blow to Children Disabled by Zika
 Luciana Magalhaes and Samantha Pearson WSJ 09/18/2020
https://www.wsj.com/articles/in-brazil-covid-19-deals-new-blow-to-children-disabled-by-zika-11600421401"

[239]"Familiares criticam MP que prevê pensão para crianças com microcefalia por zika Rodrigo Baptista Senado 10/10/2019
https://www12.senado.leg.br/noticias/materias/2019/10/10/familiares-criticam-mp-que-preve-pensao-para-criancas-com-microcefalia-por-zica"

[240]"WHO declares end of Zika emergency but says virus remains a threat; Stephanie Nebehay and Julie Steenhuysen Yahoo news 11/18/2016
https://www.yahoo.com/news/declares-end-zika-emergency-still-needs-action-192717331.html?nhp−1"

[241]"Crickets Meaning | Best 2 Definitions of Crickets .
https://www.yourdictionary.com/crickets"

[242] Op. Cit. Zika Is No Longer a Global Emergency, W.H.O. Says (Published 2016)

[243]" What Has Happened With Zika? - Blog - ISGLOBAL Elena Marbán Castro ISGlobal 19.12.2017 https://www.isglobal.org/en/healthisglobal/-/custom-blog-portlet/what-has-happened-with-zika-/5573964/0"

[244]"Zika has all but disappeared in the Americas. Why? JON COHEN Science 08/16/2017 https://www.science.org/content/article/zika-has-all-disappeared-americas-why"

[245]"Rajasthan Zika strain not linked to microcephaly The Hindu 11/03/2018 https://www.thehindu.com/sci-tech/health/rajasthan-zika-strain-not-linked-to-microcephaly/article25414057.ece"

[246] Alarm fatigue. https://en.wikipedia.org/wiki/Alarm_fatigue

[247]" Alarm fatigue: a patient safety concern. Sendelbach S, Funk M. AACN Adv Crit Care. 2013 Oct-Dec;24(4):378-86; quiz 387-8. 2013 Oct-Dec https://pubmed.ncbi.nlm.nih.gov/24153215/"

[248]Dataset Records for Secretariat of Health Surveillance Ministry of Health (Brazil) . http://ghdx.healthdata.org/organizations/secretariat-health-surveillance-ministry-health-brazil

[249] Zika Virus Infection and Associated Neurologic Disorders in Brazil | NEJM. de Oliveira, Wanderson; Christopher Dye, et al. New England Journal of Medicine 03/29/2017 https://www.nejm.org/doi/full/10.1056/nejmc1608612

[250]Journalism Is Never Perfect: The Politics of Story Corrections and Retractions JAMES MCWILLIAMS Pacific Standard 12/20/2013 https://psmag.com/social-justice/journalism-never-perfect-politics-story-corrections-retractions-71700

[251]Retractions: the good, the bad, and the ugly. What researchers stand to gain from taking more care to understand errors in the scientific record Quan-Hoang Vuong the London School of Economics 02/20/2020 https://blogs.lse.ac.uk/impactofsocialsciences/2020/02/20/retractions-the-good-the-bad-and-the-ugly-what-researchers-stand-to-gain-from-taking-more-care-to-understand-errors-in-the-scientific-record/

[252]Retraction Watch staff: Adam Marcus   . https://retractionwatch.com/meet-the-retraction-watch-staff/about-adam-marcus/

[253]Factoid | Definition of Factoid by Merriam-Webster https://www.merriam-webster.com/dictionary/factoid

[254]"Illusion of truth effect: You repeat, I believe How liars create the 'illusion of truth'
  Econowmics . http://econowmics.com/illusion-of-truth-effect-you-repeat-i-believe/"

[255]Why No One Believed Einstein Matthew Wills JSTOR Daily 08/19/2016 https://daily.jstor.org/why-no-one-believed-einstein/

CHAPTER
22. Overturning Zika's Entirely Different "Mission": Relief Through Knowledge

[256]"Medical Definition of Koch's postulates Melissa Conrad Stöppler, MD MedicineNet 03/29/2021 https://www.medicinenet.com/kochs_postulates/definition.htm"

[257]"The Scientific Method – Hypotheses, Models, Theories, and Laws
   BSC Designer Team BSC Designer 02/10/2019 https://bscdesigner.com/all-about-the-scientific-method.htm"

[258]People's Choice Awards -
https://en.wikipedia.org/wiki/People%27s_Choice_Awards

[259]Conspiracy of Interests Laurence M. Hauptman Syracuse University Press
06/21/1999
https://www.google.com/books/edition/Conspiracy_of_Interests/8gCBmfxuTO0C?hl=en

[260]"70's Ads: McDonald's You Deserve A Break Today 1971
   Youtube .
https://www.youtube.com/watch?v=0fp3co77MQw&ab_channel=PhakeNam"

[261]Gravity Is a Harsh Mistress   .
https://tvtropes.org/pmwiki/pmwiki.php/Main/GravityIsAHarshMistress